# AFTERNOTES
## goes to
# GRADUATE
# SCHOOL

# G. W. Stewart

University of Maryland
College Park, Maryland

# AFTERNOTES goes to GRADUATE SCHOOL

## Lectures on Advanced Numerical Analysis

A series of lectures on advanced numerical
analysis presented at the University of Maryland
at College Park and recorded after the fact.

Society for Industrial and Applied Mathematics

Philadelphia

**Library of Congress Cataloging-in-Publication Data**

Stewart, G. W. (Gilbert W.)
        Afternotes goes to graduate school : lectures on advanced numerical analysis : a series of lectures on advanced numerical analysis presented at the University of Maryland at College Park and recorded after the fact / G.W. Stewart.
            p.  cm.
        Includes bibliographical references and index.
        ISBN 0-89871-404-4 (pbk.)
        1. Numerical analysis. I. Title.
QA297.S784  1997
519.4--dc21                                        97-44449

# Contents

Contents

# Preface

In 1996 I published a collection of lectures entitled *Afternotes on Numerical Analysis.* The unusual name reflects the unusual way they were produced. Instead of writing the notes and giving the lectures, I gave the lectures and then wrote the notes — in real time, two lectures a week. In preparing the notes I hoped to bring the immediacy of the classroom to the printed page and give an uncluttered presentation that could be used for self-study or to supplement a course.

The original afternotes were based on an advanced undergraduate course taught at the University of Maryland. The present notes are based on the follow-up graduate course. The two courses have some overlap, but with the kind consent of my colleagues I cut out the intersection and replaced it with fresh material. The topics treated are approximation — discrete and continuous — linear and quadratic splines, eigensystems, and Krylov sequence methods. The notes conclude with two little lectures on classical iterative methods and nonlinear equations. Ordinary and partial differential equations are the subjects of yet a third course at Maryland, which may sometime inspire another set of afternotes.

I have emphasized Krylov sequence methods at the expense of classical iterative methods — an emphasis that I think is overdue. During the past quarter century the scope of Krylov algorithms has expanded until they are now the iterative methods of choice for most large matrix problems. Moreover, with these algorithms black-box code will simply not do. The user needs to know the methods to manage the programs. Consequently, I have tried to present a reasonably comprehensive survey of the basics.

The present notes are more stolid than their predecessor. Elementary numerical analysis is a marvelously exciting subject. With a little energy and algebra you can derive significant algorithms, code them up, and watch them perform. The algorithms presented here require a deeper mathematical understanding, and their implementations are not trivial. Moreover, the course and these notes assume that the student has fully mastered a one-semester introductory numerical analysis course covering the topics in the original afternotes. All this conspires to make the present notes more businesslike — and less fun. Nonetheless, I worked hard to find fresh presentations of old material and thoroughly enjoyed the process. I hope you will enjoy the results.

The notes were originally posted on the Internet, and Bo Einerson, David F. Giffiths, Eric Novak, Joseph Skudlarek, and Stephen Thomas were kind enough to send me comments and corrections. The usual suspects at SIAM — Vickie Kearn, Jean Keller-Anderson, and Corey Gray — did their usual outstanding job.

These afternotes are dedicated to Sue, Sal, and the shade of Smut, without whose persistent attentions this book would have been finished a lot sooner.

G. W. Stewart
College Park, MD

- Approximation

# Approximation

## General observations

1. Approximation problems occur whenever one needs to express a group of numbers in terms of other groups of numbers. They occur in all quantitative disciplines — engineering and the physical sciences, to name two — and they are a mainstay of statistics. There are many criteria for judging the quality of an approximation and they profoundly affect the algorithms used to solve the problem.

In spite of this diversity, most approximation problems can be stated in a common language. The best way to see this is to look at a number of different problems and extract the common features. We will consider the following here.

1. Given the radiation history of a mixture of isotopes, determine the proportion of the isotopes.
2. Determine a linear approximation to the sin function.

## Decline and fall

2. Suppose we have a mixture of two radioactive isotopes. If enough of the first isotope is present, its radiation will vary with time according to the law $ae^{-\lambda t}$. Here $a$ is a constant specifying the initial level of radiation due to the first isotope, and $\lambda$ is a decay constant that tells how fast the radiation decays. If the second isotope follows the law $be^{-\mu t}$, the mixture of the two isotopes radiates according to the law $ae^{-\lambda t} + be^{-\mu t}$.

The above law implies that the radiation will become arbitrarily small as $t$ increases. In practice, the radiation will decline until it reaches the level of the background radiation — the ambient radiation that is all around us — after which no decrease will be observed. Thus a practical model of the observed radiation $y$ from the mixture is

$$y = ae^{-\lambda t} + be^{-\mu t} + c,$$

where $c$ represents the background radiation.

3

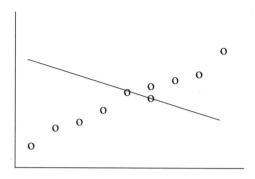

Figure 1.1. *A bad fit.*

3. Now suppose that we observe the radiation levels $y_1, y_2, \ldots, y_n$ of the mixture at times $t_1, t_2, \ldots, t_n$. Naturally, the values of the $y_i$ will be contaminated by measurement error and fluctuations in the background radiation. Can we use this dirty information to recover the initial amounts $a$ and $b$ of the two isotopes?

4. One way of proceeding is to select three of the observations — say $y_1$, $y_2$, and $y_3$ — and write down the equations

$$y_1 = ae^{-\lambda t_1} + be^{-\mu t_1} + c$$
$$y_2 = ae^{-\lambda t_2} + be^{-\mu t_2} + c$$
$$y_3 = ae^{-\lambda t_3} + be^{-\mu t_3} + c$$

This is a set of linear equations, which can be solved by standard methods for $a$, $b$, and $c$.

Although this procedure is simple, it is arbitrary, wasteful, and dangerous. It is arbitrary because it forces us to choose three of the observations. Why not choose $y_{n-2}$, $y_{n-1}$, and $y_n$ instead of $y_1$, $y_2$, and $y_3$? It is wasteful because it ignores the unchosen observations — observations that might be used to increase the accuracy of our estimates of $a$, $b$, and $c$. And it is dangerous because it can easily lead to totally wrong results.

Figure 1.1 illustrates this last point in the slightly different context of determining a line to fit a set of data. The o's, which represent the data, show a distinctly linear trend. But the unfortunate choice of two neighboring points to determine the line gives an approximation whose slope is contrary to the trend.

5. The cure for the problem is to work with all the observations at once; that is, we seek $a$, $b$, and $c$ so that

$$y_i \cong ae^{-\lambda t_i} + be^{-\mu t_i} + c, \qquad i = 1, 2, \ldots, n.$$

To make the procedure rigorous, we must associate a precise meaning with the symbol "$\cong$". We will use the principle of least squares.

Specifically, let

$$r_i = y_i - ae^{-\lambda t_i} - be^{-\mu t_i} - c, \qquad i = 1, 2, \ldots, n,$$

be the *residuals* of the approximations to the $y_i$, and let

$$\rho^2(a, b, c) = r_1^2 + r_2^2 + \cdots + r_n^2.$$

As our notation indicates, $\rho^2$ varies with $a$, $b$, and $c$. Ideally we would like to choose $a$, $b$, $c$ so that $\rho(a, b, c)$ is zero, in which case the approximation will be exact. Since that is impossible in general, the *principle of least squares* requires that we choose $a$, $b$, and $c$ so that the *residual sum of squares* $\rho^2(a, b, c)$ is minimized.

6. The principle of least squares is to some extent arbitrary, and later we will look at other ways of fitting observations. However, it has the effect of balancing the size of the residuals. If a single residual is very large, it will generally pay to reduce it. The other residuals may increase, but there will be a net decrease in the residual sum of squares.

7. Another reason for adopting the principle of least squares is computational simplicity. If we differentiate $\rho^2(a, b, c)$ with respect to $a$, $b$, and $c$ and set the results to zero, the result is a system of linear equations — commonly called the *normal equations* — that can be solved by standard techniques. But that is another story (see §6.13).

8. With an eye toward generality, let's recast the problem in new notation. Define the vectors $y$, $u$, and $v$ by

$$y = \begin{pmatrix} y_1 \\ y_2 \\ \vdots \\ y_n \end{pmatrix}, \quad u = \begin{pmatrix} e^{-\lambda t_1} \\ e^{-\lambda t_2} \\ \vdots \\ e^{-\lambda t_n} \end{pmatrix}, \quad v = \begin{pmatrix} e^{-\mu t_1} \\ e^{-\mu t_2} \\ \vdots \\ e^{-\mu t_n} \end{pmatrix},$$

and let $\mathbf{e}$ be the vector consisting of all ones. Then the vector of residuals is given by

$$r = y - au - bv - c\mathbf{e}.$$

Moreover, if for any vector $x$ we define the 2-norm of $x$ by

$$\|x\|_2 = \sqrt{\sum_i x_i^2},$$

then the principle of least squares can be cast in the form:

> Choose $a$, $b$, and $c$ so that $\|y - au - bv - c\mathbf{e}\|_2$ is minimized.

In this notation we see that the natural setting for this kind of least squares approximation is the space $\mathbf{R}^n$ of real $n$-dimensional vectors equipped with the 2-norm.

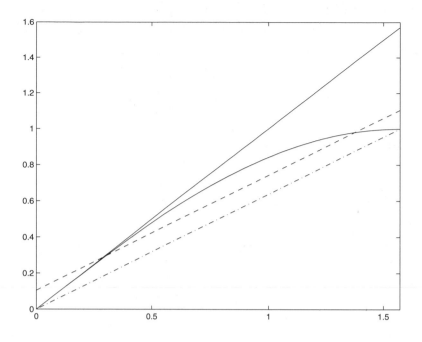

Figure 1.2. *Approximations to* $\sin t$.

## The linear sine

9. An important problem in numerical analysis is the computation of special functions like the trigonometric or hyperbolic functions. Here we will consider a very simple problem: Approximate $\sin x$ on the interval $[0, \pi/2]$ by a straight line $\ell(t) = at + b$.

10. A natural approach is to use a Taylor series expansion. Specifically, we can write

$$\sin t = t - \frac{t^3}{3!} + \frac{t^5}{5!} - \cdots .$$

Truncating this expansion at the linear term, we obtain our first approximation to $\sin t$:

$$\sin t \cong \ell_1(t) \equiv t.$$

The approximation is represented by the solid line in Figure 1.2. Since the Taylor expansion is taken about zero, the approximation is very good for small $t$. For $t = \pi/2$, however, the error is about 0.57, which most people would consider unacceptably large.

11. Another possibility is to chose the line passing through the endpoints of the graph of $\sin t$ (the dot-dash line in Figure 1.2). This approximation is given

by

$$\ell_2(t) = \frac{2t}{\pi}.$$

The maximum error in this approximation is about 0.211 and is attained at the point $t = 0.881$ — a considerable improvement over the Taylor series expansion.

12.   However, we can do better. Imagine that we move the dotted line in Figure 1.2 upward while preserving its slope. The maximum error at $t = 0.881$ will decrease. The error at the endpoints will increase; but until the line moves half the original maximum error, the error at the endpoints will be strictly less than the error at $t = 0.881$. Thus, if we set

$$\ell_3 = 0.105 + \frac{2t}{\pi}$$

(note that $0.105 \cong 0.211/2$), we obtain an approximation whose maximum error occurs at three points, as shown in the following table.

$$\begin{array}{cccc} t & : & 0.000 \quad 0.881 \quad 1.000 \\ \sin t - \ell_3(t) : & -0.105 \quad 0.105 \quad -0.105 \end{array} \tag{1.1}$$

This approximation is illustrated by the dashed line in Figure 1.2.

13.   This is as good as it gets. We will see later that any attempt to change the coefficients of $\ell_3$ must result in an error larger than 0.105 somewhere on the interval $[0, \pi/2]$. Thus we have found the solution of the following problem. Choose $a$ and $b$ so that $\max_{t \in [0, \pi/2]} |\sin t - a - bt|$ is minimized.[1]

14.   Once again it will be helpful to introduce some new notation. Let $\|\cdot\|_\infty$ be defined for all continuous functions on $[0, \pi/2]$ by

$$\|f\|_\infty = \max_{t \in [0, \pi/2]} |f(t)|.$$

Then our approximation problem can be stated in the following form.

> Choose $a$ and $b$ so that $\|\sin t - a - bt\|_\infty$ is minimized.

## Approximation in normed linear spaces

15.   We are now in a position to set down the common elements of our problems. They are two in number.

1.   A vector space, also called a linear space.

---

[1]This example also shows that it is possible to focus so closely on minimizing error that other desirable properties are lost. Our linear sine does not satisfy $t^{-1} \sin t \to 0$ as $t \to 0$ (see §4.17).

2.  A norm.

We will look at each in turn.

16. Informally, a real vector space is a set of objects, called vectors or elements, that can be added, subtracted, and multiplied by real numbers (also called scalars). The usual rules of commutativity, associativity, and distributivity hold for these operations.

The vector space underlying our first problem is the space $\mathbf{R}^n$ of all column vectors of the form

$$\begin{pmatrix} x_1 \\ x_2 \\ \vdots \\ x_n \end{pmatrix}.$$

This space — or its complex analogue — is the space of applied linear algebra and matrix computations.

The space underlying our second problem is the set

$$C[0, \pi/2] = \{f : f \text{ is continuous on } [0, \pi/2]\}.$$

To people reared in finite-dimensional spaces, it may seem odd to have functions as elements of a vector space. But they certainly can be added, subtracted, and multiplied by scalars.

17.    A norm on a vector space is a function $\|\cdot\|$ satisfying the following conditions.

1.  $x \neq 0 \implies \|x\| > 0$ (definiteness).
2.  If $\alpha$ is a scalar, then $\|\alpha x\| = |\alpha|\|x\|$ (homogeneity).
3.  $\|x + y\| \leq \|x\| + \|y\|$ (triangle inequality).

Norms are generalizations of the absolute value of a number. Their introduction allows many results from analysis on the real line to be transferred to more general settings.

18. A *normed linear space* is a pair consisting of a vector space $\mathcal{V}$ and a norm $\|\cdot\|$ on $\mathcal{V}$. We will be concerned with the following linear approximation problem.

> Given a normed vector space $(\mathcal{V}, \|\cdot\|)$ and elements $y, v_1, \ldots, v_k \in \mathcal{V}$, determine constants $\beta_1, \ldots, \beta_k$ so that
>
> $$\|y - \beta_1 v_1 - \cdots - \beta_k v_k\| \text{ is minimized.} \qquad (1.2)$$
>
> We call such an approximation a *best approximation*.

Both our problems fit neatly into this mold of best approximation.

## Significant differences

19. It might be thought that we could now go on to develop a general theory of approximation in normed linear spaces, and that would be the end of the topic. Certainly we can say may things in general; and when we can, the place to say them is in normed linear spaces. However, normed linear spaces vary greatly in their properties, and these variations force us to treat them individually.

20. A striking difference between $\mathbf{R}^n$ and $C[0,1]$ (note the change of interval) is that the former is finite dimensional while the latter is infinite dimensional. The space $\mathbf{R}^n$ can have no more than $n$ linearly independent vectors. Since dependent vectors are redundant, the finite dimensionality of $\mathbf{R}^n$ effectively limits the number $k$ in (1.2) to be not greater than $n$, at least if we want a unique solution.

On the other hand, the monomials $1, t, t^2, \ldots$ are independent — no finite, nontrivial linear combination of them can be identically zero. Consequently, if we identify $v_i$ with $t^{i-1}$ in (1.2) we can take $k$ as large as we wish. This raises a question that does not arise in a finite-dimensional setting. If we denote by $p_k$ the polynomial of order $k$ that best approximates $y$, does $p_k$ converge to $y$ as $k \rightarrow \infty$? We will return to this problem in §3.9.

21. Since a normed linear space consists of both a vector space and a norm, changing the norm changes the normed linear space even when the underlying vector space remains the same. For example, the analogue of the $\infty$-norm on $\mathbf{R}^n$ is defined by

$$\|x\|_\infty = \max_i |x_i|.$$

The approximation problem in the $\infty$-norm will generally have a different solution than the approximation problem in the 2-norm. Moreover, the algorithms that compute solutions for the two cases are quite different. The algorithms for approximation in the $\infty$-norm are more complicated because $\|x\|_\infty$ is not a smooth function of $x$.

22. In infinite-dimensional settings, the choice of norms often determines the space itself. For example, the natural generalization of the 2-norm to spaces of functions on $[0,1]$ is defined by

$$\|f\|_2^2 = \int_0^1 f(t)^2 \, dt,$$

and the natural underlying space is the set of all functions $f$ for which $\|f\|_2 < \infty$. This set is richer than $C[0,1]$. For example, $t^{-\frac{1}{3}}$ is not continuous — it asymptotes to infinity at zero. But $\|t^{-\frac{1}{3}}\|_2 = \sqrt{3}$.

# Approximation

The Space $C[0,1]$
Existence of Best Approximations
Uniqueness of Best Approximations
Convergence in $C[0,1]$
The Weierstrass Approximation Theorem
Bernstein Polynomials
Comments

## The space $C[0,1]$

1. Since normed linear spaces differ in their properties, we will treat each on its own terms, always taking care to point out results that hold generally. We will begin with $C[0,1]$ equipped with the $\infty$-norm. (Unless another norm is specified, $C[0,1]$ comes bundled with the $\infty$-norm.)

2. In investigating $C[0,1]$, the first order of business is to verify that the $\infty$-norm is well defined. Here the fact that the interval $[0,1]$ is closed is critical. For example, the function $x^{-1}$ is continuous on the half-open interval $(0,1]$. But it is unbounded there, and hence has no $\infty$-norm.

   Fortunately, a theorem from elementary analysis says that a function continuous on a finite closed interval attains a maximum on that set. In other words, if $f \in C[0,1]$, not only is $\|f\|_\infty$ well defined, but there is a point $x_*$ such that

$$\|f\|_\infty = f(x_*).$$

3. The fact that $[0,1]$ is closed and bounded has another important consequence for the elements of $C[0,1]$ — they are uniformly continuous. The formal definition of uniform continuity is a typical piece of epsilonics. A function $f$ is uniformly continuous on $[0,1]$ if for every $\epsilon > 0$ there is a $\delta > 0$ such that

$$x, y \in [0,1] \quad \text{and} \quad |x - y| < \delta \implies |f(x) - f(y)| < \epsilon.$$

The common sense of this definition is that we can keep the function values near each other by keeping their arguments near each other and the process does not depend on where we are in $[0,1]$.

   Like many definitions of this kind, uniform continuity is best understood by considering counterexamples. You should convince yourself that $x^{-1}$ is not uniformly continuous on $(0,1]$.

## Existence of best approximations

4.  Another problem we must investigate before proceeding further is whether the approximation problem of minimizing

$$\|y - \beta_1 v_1 - \cdots - \beta_k v_k\|$$

[cf. (1.2)] has a solution. In fact it does — and in any normed linear space, not just $C[0,1]$.

    The proof, which we omit, is made rather easy because we have hedged in defining our problem. All the action is in the finite-dimensional subspace spanned by $y, v_1, \ldots, v_k$. If we had looked for approximations in an infinite-dimensional subspace, we would have had to worry about whether the subspace is closed in some appropriate sense. Fortunately, most real-life approximation problems are of our restricted form.

## Uniqueness of best approximations

5.  If the elements $v_i$ are linearly dependent, the best approximation is not unique. For if $\beta_1 v_1 + \cdots + \beta_k v_k$ is a solution and if there is a nontrivial linear combination of the $v_j$ that is zero — say, $\alpha_1 v_1 + \cdots + \alpha_k v_k = 0$ — then $(\alpha_1 + \beta_1)v_1 + \cdots + (\alpha_k + \beta_k)v_k$ is a different solution.

6.  Even when the $v_i$ are independent, the solution may not be unique. We will see later that least squares solutions are unique. But in $C[0,1]$ the problem of approximating 1 by $(t - \frac{1}{2})^2$ has the continuum of solutions

$$a\left(t - \frac{1}{2}\right)^2, \qquad a \in [0,8].$$

In general the problem of uniqueness is difficult. It depends on the norm and, in some instances, the particular problem.

## Convergence in $C[0,1]$

7.  There is a natural way to define convergence in a normed linear space. Let $v_1, v_2, \ldots$ be an infinite sequence of elements. Then $v_i$ *converges in norm* to $v$ if

$$\lim_{k \to \infty} \|v_k - v\| = 0.$$

Such convergence is also called *strong convergence*.

8.  The main problem with convergence in norm is that the underlying vector space may have a structure all its own that generates a different notion of convergence. For example, it is natural to say that a vector in $\mathbf{R}^n$ converges if its components converge. It is a remarkable fact that this kind of convergence is equivalent to convergence in norm for any norm whatsoever. As far as convergence is concerned, all norms in $\mathbf{R}^n$ are equivalent.

9.   The elements of $C[0,1]$ are functions for which a natural notion of convergence also exists. Specifically, a sequence of functions $f_1, f_2, \ldots$ on $[0,1]$ *converges pointwise* to a function $f$ if

$$\lim_{k \to \infty} f_k(t) = f(t), \qquad t \in [0,1].$$

However, for $C[0,1]$ pointwise convergence is not equivalent to convergence in norm. For example, consider the functions

$$f_k(t) = \sin \pi t^k.$$

It is easily seen that these functions converge pointwise to zero on $[0,1]$. But since the maximum value of $f_k(t)$ in $[0,1]$ is one, we have $\|f_k - 0\|_\infty = 1$, so that the $f_k$ do not converge to zero in norm.

10.   The pointwise equivalent of convergence in norm in $C[0,1]$ is *uniform convergence*. A sequence $f_1, f_2, \ldots$ on $[0,1]$ converges uniformly to $f$ if for every $\epsilon > 0$ there is a $K$ such that

$$k > K \implies |f_k(t) - f(t)| < \epsilon, \qquad t \in [0,1]. \tag{2.1}$$

In other words, the values of $f_k$ not only get near the corresponding values of $f$, but they do it uniformly across the entire interval $[0,1]$. Since we can rewrite (2.1) in the form

$$k > K \implies \|f_k - f\|_\infty < \epsilon, \qquad t \in [0,1],$$

uniform convergence is exactly the same as convergence in the $\infty$-norm.

11.   Uniform convergence has important consequences. For our purposes the chief one is that the uniform limit of continuous functions on $[0,1]$ is continuous. Since convergence in the infinity norm is equivalent to uniform convergence, this means that functions in $C[0,1]$ cannot converge in norm to discontinuous functions — the space $C[0,1]$ is closed under uniform limits.

## The Weierstrass approximation theorem

12.   Every finite-dimensional vector space has a basis. That is, there are independent vectors $b_1, \ldots, b_n$ such that every vector $v$ in the space can be expressed uniquely in the form

$$v = \alpha_1 b_1 + \cdots + \alpha_n b_n.$$

The number $n$ is unique and is called the dimension of the space.

   In infinite-dimensional spaces like $C[0,1]$ the concept of basis becomes tricky. For example, since any basis must necessarily be infinite, what does it mean for the basis elements to be independent? In many applications, however, it is sufficient to find a set of elements $\{v_1, v_2, \ldots\}$ whose finite linear combinations can approximate any element of the space to arbitrary accuracy. Such a set is called a *complete basis*.

13. In turns out that the set $\{1, t, t^2, \ldots\}$ is complete in $C[0,1]$—a result due to Weierstrass. As with most problems involving sums of powers, the result is best cast in terms of polynomials. Specifically, a finite linear combination of powers of $t$ has the form

$$p(t) = a_0 + a_1 t + \cdots + a_k t^k;$$

i.e., it is a polynomial. Thus to say that the set $\{1, t, t^2, \ldots\}$ is complete is to say that every member of $C[0,1]$ can be approximated to arbitrary accuracy by a polynomial. This is the usual form of Weierstrass's theorem.

> Let $f \in C[0,1]$. Then for every $\epsilon > 0$ there is a polynomial $p$ such that
> $$\|p - f\|_\infty \le \epsilon.$$

## Bernstein polynomials

14. The amazing thing about this result is that we can write down a simple formula for a sequence of polynomials that converges to $f$. The construction is due to the Russian mathematician Bernstein. He derived his polynomials by probabilistic reasoning, and it is worthwhile to retrace his steps. We begin with a rather silly way of approximating $f(\frac{1}{2})$.

15. We start by choosing an integer $k$ and flipping a coin $k$ times. We then count the number of heads and divide by $k$ to get a number $s \in [0,1]$. We then evaluate $f(s)$ and that is our approximation.

At first sight, this process has nothing to recommend it. The number $s$ is a random variable that can assume the values

$$t_{i,k} = \frac{i}{k}, \qquad i = 0, 1, \ldots, k.$$

When $k = 1$ these values are 0 and 1. Thus our procedure returns either $f(0)$ or $f(1)$, each with probability $\frac{1}{2}$. This does not tell much about $f(\frac{1}{2})$.

However, things look better as $k$ increases. Figure 2.1 shows the probability distribution of the $t_{i,k}$ for $k = 100$ (not to scale) together with a plot of a representative function. Note that it is improbable that $s$ will lie outside the interval $[0.4, 0.6]$. Since $f(t)$ varies slowly in that interval, it is probable that we will get a reasonable approximation to $f(\frac{1}{2})$.

16. To evaluate $f$ at an arbitrary point $t$ in $[0,1]$ we use a biased coin that comes up heads with probability $t$. The prescription for choosing a point $s$ at

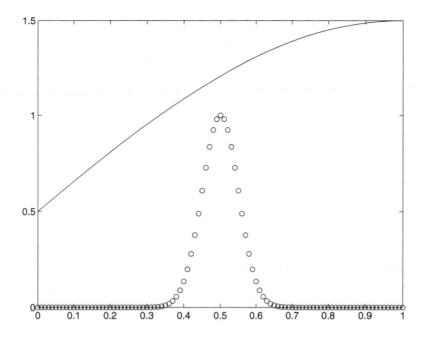

Figure 2.1. *Distribution of s.*

which to evaluate $f$ remains unchanged. But because our coin is biased, the probability that $s = t_{i,k}$ changes. Specifically,

the probability that $s = t_{i,k}$ is $a_{i,k}(t) = \binom{k}{i} t^i (1-t)^{k-i}$.

The construction of the sequence of polynomials depends on the properties of the $a_{i,k}(t)$. Because we must evaluate $f$ at some $t_{i,k}$, we must have

$$\sum_{i=0}^{k} a_{i,k}(t) = 1. \qquad (2.2)$$

The mean value of $s$ is

$$\sum_{i=0}^{k} a_{i,k}(t) t_{i,k} = t, \qquad (2.3)$$

and its variance is

$$\sum_{i=0}^{k} a_{i,k}(t)(t_{i,k} - t)^2 = \frac{t(1-t)}{k}. \qquad (2.4)$$

Since the variance goes to zero as $k$ increases, the distribution of $s$ bunches up around $t$ and we can expect to evaluate $f$ near $t$. In particular, we would expect the average value of $f(s)$ to be near $f(t)$ and to get nearer as $k$ increases. Now this average value is given by

$$B_k(t) = \sum_{i=0}^{k} a_{i,k}(t)f(t_{i,k}) = \sum_{i=0}^{k} f(t_{i,k})\binom{k}{i}t^i(1-t)^{k-1}, \qquad (2.5)$$

which is a polynomial of degree not greater than $k$ in $t$. The polynomials $B_k$ are called the *Bernstein polynomials associated with* $f$.

17. It can be shown that the polynomials $B_k$ converge uniformly to $f$. The proof is essentially an analytic recasting of the informal reasoning we used to derive the polynomials, and there is no point in repeating it here. Instead we will make some comments on the result itself.

## Comments

18. Although we have stated the approximation theorem for the interval $[0,1]$, it holds for any closed interval $[a,b]$. For the change of variable

$$t = a(1-s) + bs$$

transforms $[a,b]$ (in $t$) to $[0,1]$ (in $s$).

19.   Bernstein's approach to the Weierstrass approximation theorem is constructive, but it should not be thought that his polynomials are the way to go in approximating functions. For one thing, the Bernstein polynomial of degree $k$ that approximates a polynomial of degree $k$ is not necessarily the polynomial itself. Moreover, the convergence can be very slow. For example, if we expand the right-hand side of (2.4), we get

$$\sum_{i=1}^{k} a_{i,k}(t)t_{i,k}^2 - 2\sum_{i=1}^{k} a_{i,k}(t)t + \sum_{i=1}^{k} a_{i,k}(t)t^2 = \frac{t(1-t)}{k}.$$

If we take $f(t) = t^2$, it follows from (2.5), (2.2), and (2.3) that

$$B_k(t) - f(t) = \frac{t(1-t)}{k} \le \frac{1}{4k}.$$

In this case, to reduce the maximum error in the approximation below $10^{-6}$, we would have to take $k > 250,000$.

20. In spite of the slow convergence of the Bernstein polynomials, the Weierstrass theorem is quite valuable. It allows us to deduce the convergence of a number of important numerical procedures, including best uniform approximation (§3.9) and Gaussian quadrature. It does not give the rate of convergence, but knowing that a process converges at all is often a help in establishing how fast it converges.

# Approximation

Chebyshev Approximation
Uniqueness
Convergence of Chebyshev Approximations
Rates of Convergence: Jackson's Theorem

## Chebyshev approximation

1. We now turn to the problem of finding a polynomial of degree $k$ that best approximates a continuous function $f$ in $C_\infty[0,1]$. This problem is known as *Chebyshev approximation,* after the Russian mathematician who first considered it. It is also known as *best uniform approximation.* Our first order of business will be to establish a characterization of the solution.

2. The linear approximation to the sine function that we derived in §§1.9–1.14 is prototypical. The table in (1.1) shows that the maximum error is attained at three points and the errors at those points alternate in sign. It turns out that the general Chebyshev approximation of degree $k$ will assume its maximum error at $k+2$ points and the errors will alternate in sign.

3. We now wish to establish the following characterization of the Chebyshev approximation.

Let $f \in C[0,1]$ and let $p_k$ be of degree $k$. Let

$$e(t) = f(t) - p_k(t).$$

Then a necessary and sufficient condition that $p_k$ be a Chebyshev approximation of $f$ is that there exist $k+2$ points

$$0 \le t_0 < t_1 < \cdots < t_k < t_{k+1} \le 1$$

such that

1. $|e(t_i)| = \|f - p_k\|_\infty, \qquad i = 0, \ldots, k+1,$
2. $e(t_i)e(t_{i+1}) \le 0, \qquad i = 0, \ldots, k.$

(3.1)

The second condition in (3.1) says that successive errors $e(t_i)$, if they are not zero, alternate in sign. For brevity we will say that the maximum error alternates at $k+2$ points.

4. The proof of this result consists of denying the characterization and showing that we can then get a better approximation. Before launching into the

17

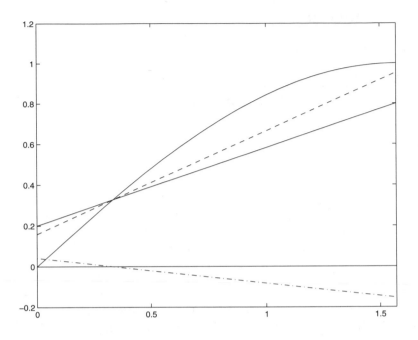

Figure 3.1. *Improving an approximation to the sin.*

proof, which is complicated, we consider a simple example. Figure 3.1 shows an approximation $p(t)$ to $\sin t$ (solid line) that does not satisfy the characterization — its maximum error alternates in sign only at the endpoints of the interval $[0, \pi/2]$. Then there must be a point $t_0$ ($\cong 0.3336$) between the endpoints at which the error is zero. We can pass a line $q(t)$ through $t_0$ that is small and has the same sign as the error at the endpoints (dot-dash line). Then $p(t) - q(t)$ (dashed line) is a better approximation to the sine.

In the general proof of necessity, we take an approximation that does not satisfy the conditions of the theorem and choose a subset of the points where it crosses the function — i.e., where the error is zero. We then construct a polynomial whose roots are these points and show that by adding a suitable multiple of it we can improve the approximation. The proof consists of a largely geometric construction of the polynomial and a largely analytic verification that the polynomial does the job.

5. Turning now to the general proof of the necessity of the condition, suppose that the maximum error — call it $\epsilon$ — does not alternate at $k + 2$ points. We

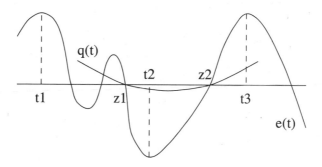

Figure 3.2. *The points $t_i$ and $z_i$.*

will construct points $t_i$ as follows.

$t_1$   is the first point in $[0, 1]$ for which $|e(t_1)| = \epsilon$.

$t_2$   is the first point greater than $t_1$ with $|e(t_2)| = \epsilon$ and $e(t_1)e(t_2) < 0$.

$t_3$   is the first point greater than $t_2$ with $|e(t_3)| = \epsilon$ and $e(t_2)e(t_3) < 0$.

etc.

This process must terminate with $\ell$ points $t_i$, where $\ell < k + 2$. This situation is illustrate for $\ell = 3$ in Figure 3.2.

Now since $e(t_i)$ and $e(t_{i+1})$ $(i = 1, \ldots, \ell - 1)$ have opposite signs, there is a zero of $e$ between them. Let $z_i$ be the largest such zero. In addition let $z_0 = 0$ and $z_\ell = 1$.

Let
$$q(t) = (t - z_1)(t - z_2) \cdots (t - z_{\ell-1}).$$

Since $\ell < k + 2$, the degree of $q$ is no greater than $k$. Consequently, for any $\lambda > 0$ the polynomial $p_k - \lambda q$ is a candidate for a best approximation. We are going to show that for $\lambda$ small enough the error

$$(p_k - \lambda q) - f = e - \lambda q$$

has norm strictly smaller than $\epsilon$. This implies that $p_k$ was not a best approximation in the first place. We will assume without loss of generality that

$$\text{sign}[q(t)] = \text{sign}[e(t_i)], \qquad t \in (z_{i-1}, z_i).$$

We will first consider the determination of $\lambda$ for the interval $[z_0, z_1]$. For definiteness suppose that $\text{sign}[e(t_i)] > 0$ (which corresponds to Figure 3.2). This implies that for any $\lambda > 0$

$$\epsilon > e(t) - \lambda q(t), \qquad t \in (z_{i-1}, z_i).$$

What we must show now is that for any sufficiently small $\lambda > 0$

$$e(t) - \lambda q(t) > -\epsilon, \qquad t \in (z_{i-1}, z_i).$$

Let

$$\eta_0 = \min_{t \in [z_0, z_1]} \{e(t)\}.$$

We must have $\epsilon + \eta_0 > 0$. For otherwise the interval $[z_0, z_1]$ must contain a point with $e(t) = -\epsilon$, contradicting the construction of $t_2$. Now let

$$\lambda_0 = \frac{\epsilon + \eta_0}{2\|q\|_\infty}.$$

Then for $0 \le \lambda \le \lambda_0$ and $t \in [z_0, z_1]$,

$$
\begin{aligned}
e(t) - \lambda q(t) &\ge e(t) - \lambda_0 q(t) \\
&\ge \eta_0 - \frac{(\epsilon+\eta_0)q(t)}{2\|q\|_\infty} \\
&\ge \eta_0 - \frac{\epsilon+\eta_0}{2} \\
&\ge -\frac{\epsilon-\eta_0}{2} \\
&> -\epsilon \qquad\qquad \text{(since } \eta_0 > -\epsilon\text{),}
\end{aligned}
$$

which is the inequality we wanted.

Parameters $\lambda_i$ corresponding to the interval $[z_{i-1}, z_i]$ are determined similarly. If $0 < \lambda < \min\{\lambda_i\}$, then $\|f - (p - \lambda q)\|_\infty < \epsilon$, which establishes the contradiction.

6. Turning now to sufficiency, suppose that the maximum error in $p_k$ alternates at the points $t_0, \ldots, t_{k+1}$ but that $p_k$ is not a Chebyshev approximation. Let $q$ be a Chebyshev approximation. Denote the error in $p$ and $q$ by $e_p$ and $e_q$. Since $|e_q(t_i)| < |e_p(t_i)|$, the values of $e_p - e_q$ at the $t_i$ alternate in sign. But $e_p - e_q = q - p_k$, whose values also alternate at the $t_i$. It follows that $q - p_k$ has $k + 1$ zeros between the $t_i$. But $q - p_k$ has degree not greater than $k$. Hence $q - p_k = 0$, or $p_k = q$, a contradiction.

## Uniqueness

7. The characterization of best uniform approximations does not insure their uniqueness. It is conceivable that there could be two best approximations that alternate at a different set of points. Fortunately, it is easy to show that this cannot happen.

8. Let $p$ and $q$ be best approximations of degree $k$ of $f$, and set

$$\epsilon = \|p - f\|_\infty = \|q - f\|_\infty.$$

We claim that $\frac{1}{2}(p+q)$ is also a best approximation. In fact,

$$\left\|\frac{1}{2}(p+q) - f\right\|_\infty \le \frac{1}{2}\|p - f\|_\infty + \frac{1}{2}\|q - f\|_\infty = \epsilon.$$

Since we cannot have $\|\frac{1}{2}(p+q) - f\|_\infty < \epsilon$, we must have $\|\frac{1}{2}(p+q) - f\|_\infty = \epsilon.$[2]

By the characterization of best approximations, there must be points $t_i$ $(i = 0, 1, \ldots, k+1)$ such that (say)

$$\frac{1}{2}[p(t_i + q(t_i))] - f(t_i) = \frac{1}{2}[p(t_i) - f(t_i)] + \frac{1}{2}[q(t_i) - f(t_i)] = (-1)^i\epsilon. \quad (3.2)$$

Since $|p(t_i) - f(t_i)| \le \epsilon$ and $|q(t_i) - f(t_i)| \le \epsilon$, the only way for (3.2) to hold is for

$$p(t_i) - f(t_i) = q(t_i) - f(t_i) = (-1)^i\epsilon.$$

It follows that $p(t_i) = q(t_i)$. Thus $p$ and $q$ are polynomials of degree $k$ that are equal at $k + 2$ distinct points. Hence $p = q$.

## Convergence of Chebyshev approximations

9. We have seen that for each function $f \in C_\infty[0, 1]$ there is a sequence of best approximations $p_0, p_1, \ldots$, where each $p_k$ is of degree not greater than $k$. It is natural to conjecture that this sequence actually converges to $f$ in norm. The proof is trivial.

10. Let $B_k(f)$ be the $k$th Bernstein polynomial corresponding to $f$ (see §2.16). Our proof of the Weierstrass approximation theorem shows that $\|B_k(f) - f\|_\infty \to 0$. Since $p_k$ is a best approximation,

$$\|p_k - f\|_\infty \le \|B_k(f) - f\|_\infty \to 0.$$

Hence $\|p_k - f\|_\infty \to 0$.

## Rates of convergence: Jackson's theorem

11. This proof of the convergence of best approximations shows only that the convergence is no slower than the convergence of the Bernstein polynomials. That is not saying much, since the Bernstein polynomials are known to converge slowly (see §2.19). A stronger result goes under the name of Jackson's theorem.[3]

---

[2] More generally any convex combination — that is a combination of the form $\lambda p + (1-\lambda)q$ $(\lambda \in [0, 1])$ — is of best approximations is a best approximation, and this result holds in any normed linear space.

[3] Jackson established the basic results in 1911, results that have since been elaborated. The theorem stated here is one of the elaborations. For more, see E. W. Cheney, *Introduction to Approximation Theory*, McGraw–Hill, New York, 1966.

> Let $f$ have $n$ continuous derivatives on $[0,1]$, and let $p_k$ be the best approximation to $f$ of degree $k$. Then there is a constant $\kappa_n$, independent of $f$ and $k$, such that
> $$\|p_k - f\|_\infty \le \frac{\kappa_n \|f^{(n)}\|_\infty}{k^n}.$$

12. A little notation will be helpful here. Let $p_k$ be a Chebyshev approximation of degree $k$ to $f$ on $[0,1]$. Then we will write

$$E_k(f) = \|p_k - f\|_\infty$$

for the maximum error in $p_k$. With this notation Jackson's theorem can be written in the form

$$E_k(f) \le O(k^{-n}). \tag{3.3}$$

This kind of convergence, although it can be initially swift, tends to stagnate. For example, if $n = 2$, then when $k = 10$, the right-hand side of (3.3) is reduced by a factor of 100. But a further increment of 10 ($k = 20$) only decreases it by a factor of four. Increasing $k$ from 20 to 30 only decreases it by a factor of 0.44.

However, if $n$ is at all large, the initial decrease may be sufficient. For example, if $n = 10$, the right-hand side of (3.3) is proportional to $10^{-10}$.

13. It might be thought that if $f$ is infinitely differentiable then Jackson's theorem would imply infinitely fast convergence. However, the constant $\kappa_n$ grows with $n$. Consequently, it takes longer and longer for the convergence predicted by Jackson's theorem to set in. Nonetheless, the convergence of best approximations is impressive, as we shall see later.

# Approximation

A Theorem of de la Vallée Poussin
A General Approximation Strategy
Chebyshev Polynomials
Economization of Power Series
Farewell to $C[a, b]$

## A theorem of de la Vallée Poussin

1. We now turn to methods for calculating the best approximation to a function $f$. There is a method, called the Remes algorithm, that will compute a best approximation to high accuracy.[4] However, high accuracy is seldom needed in best polynomial approximation. This is a consequence of the following theorem of de la Vallée Poussin.

---

Let $p$ be a polynomial of degree not greater than $k$ such that $p - f$ alternates at the points $t_0 < t_1 < \cdots < t_{k+1}$; i.e., $\mathrm{sign}[p(t_{i+1}) - f(t_{i+1})] = -\mathrm{sign}[p(t_i) - f(t_i)]$ $(i = 0, \ldots, k+1)$. Then

$$\min_i |p(t_i) - f(t_i)| \le E_k(f) \le \|p - f\|_\infty. \qquad (4.1)$$

---

The right-hand inequality in (4.1) is simply a statement that the error in any polynomial approximation is not less than the error in the best polynomial approximation. The left-hand inequality is established exactly as we established the sufficiency of the characterization of best approximations in §3.6. Assume that the best approximation $p_k$ has error less than $\min_i |p(t_i) - f(t_i)|$ and show that $p_k = p$.

2. The significance of this result can be more easily seen by writing

$$\rho^{-1} \equiv \frac{\min_i |p(t_i) - f(t_i)|}{\|p - f\|_\infty} \le \frac{E_k(f)}{\|p - f\|_\infty}$$

so that

$$\|p - f\|_\infty \le \rho E_k(f).$$

We will call the number $\rho$ the *V-P ratio for the polynomial $p$ at the points $t_i$.* If it is near one, the polynomial $p$ has a maximum error that is almost as good as the best approximation. Even when $\rho$ is equal to, say, 2, the maximum error in polynomial $p$ is no more than twice the error in the best approximation. If

---

[4]A FORTRAN program is available on Netlib (http://www.netlib.org/index.html).

23

we are talking about errors of order, say, $10^{-5}$, then a factor of two is not very much.

It is important to stress that the V-P ratio is defined entirely in terms of the $f$ and $p$ and can be computed — at least to reasonable accuracy. One way is to evaluate $f$ and $p$ over a fine grid of equally spaced points and use the values to approximate $\|f - p\|_\infty$ and to determine good points of alteration $t_i$.

## A general approximation strategy

3. The above considerations suggest that instead of trying to compute the best approximation we look for quick-and-dirty methods to obtain approximations that alternate with a reasonable V-P ratio. Here is how we might go about it.

Suppose we have a sequence of polynomials $c_k$ that equi-alternate on the interval under consideration. In other words, there are points $t_{0,k}, t_{1,k}, \ldots, t_{k,k}$ such that

$$c_k(t_{i+1,k}) = -c_k(t_{i,k}), \qquad i = 0, 1, \ldots, k - 1.$$

Suppose further that we can expand our function in terms of these polynomials, so that

$$f(t) = b_0 c_0(t) + b_1 c_1(t) + b_2 c_2(t) + \cdots. \tag{4.2}$$

If we truncate this expansion at $c_k$, then the remainder will be

$$e(t) = b_{k+1} c_{k+1}(t) + b_{k+2} c_{k+2}(t) + \cdots.$$

If the $b_i$ decrease quickly enough, the error will be dominated by the term $b_{k+1} c_{k+1}(t)$, which equi-alternates; i.e., it has a V-P ratio of one. The effect of the other terms is to increase the V-P ratio. But, as we have observed, we can increase the ratio and still have a satisfactory approximation.

4. The strategy just sketched leaves open two questions.

1. Do the polynomials $c_n$ exist, and if so, how are they constructed?

2. How do we get expansions like (4.2)?

The answer to the first question is that the required polynomials exist. They are called Chebyshev polynomials. Chebyshev polynomials are not only important in approximation theory, but they pop up in other areas of numerical analysis. We will treat them in the next division.

The answer to the second question is that there are a variety of ways to obtain the required expansions. We will consider an extremely simple one called economization of power series.

## Chebyshev polynomials

5. The equi-alternating *Chebyshev polynomials*[5] are defined in terms of trigonometric and hyperbolic functions:

$$c_k(t) = \begin{cases} \cos(k\cos^{-1} t), & |t| \le 1, \\ \cosh(k\cosh^{-1} t), & |t| \ge 1. \end{cases} \quad (4.3)$$

(Actually, both formulas are valid for all values of $t$ but can give complex intermediate quantities if they are used outside their natural ranges.)

6. To see that these functions are actually polynomials, we will show that they satisfy a simple recurrence. To get us started, note that $p_0(t) = 1$ and $p_1(t) = t$.

Now let $t = \cos\theta$, so that $p_k(t) = \cos k\theta$. By adding the trigonometric identities

$$\cos(k+1)\theta = \cos k\theta \cos\theta - \sin k\theta \sin\theta,$$
$$\cos(k-1)\theta = \cos k\theta \cos\theta + \sin k\theta \sin\theta, \quad (4.4)$$

we get

$$\cos(k+1)\theta + \cos(k-1)\theta = 2\cos k\theta \cos\theta$$

or in terms of Chebyshev polynomials

$$c_{k+1}(t) = 2tc_k(t) - c_{k-1}(t). \quad (4.5)$$

From this it is obvious that $c_k$ is a polynomial of degree $k$ and that its leading term is $2^{k-1}t^k$.

7. Since the hyperbolic functions satisfy the identities (4.4), the same argument shows the identity of the two expressions in (4.3).

8. Since $p_k(t) = \cos(k\cos^{-1} t)$ on $[-1,1]$, the absolute value of $p_k(t)$ cannot be greater than one on that interval. If we set

$$t_{i,k} = \cos\frac{(k-i)\pi}{k}, \qquad i = 0,1,\ldots,k,$$

then

$$c_k(t_{i,k}) = \cos(k-i)\pi = (-1)^{k-i}.$$

Thus the values of $c_k(t)$ alternate between plus and minus one as $t$ varies from $-1$ to 1. The points $t_{i,k}$ are called the *Chebyshev points* of $c_k$.

9. If it is desired to evaluate $c_k$ at the point $t$, we can start with $c_0(t) = 1$ and $c_1(t) = t$ and apply the recurrence (4.5). This procedure has the advantage

[5]Chebyshev polynomials are frequently written $T_k(t)$, which reflects an alternate transcription — Tchebycheff — from the Cyrillic alphabet.

that it produces values for all the polynomials of degree less than or equal to $k$. Consequently, we can also evaluate sums of the form

$$b_0 c_0(t) + b_1 c_1(t) + \cdots + b_k c_k(t)$$

in $O(k)$ operations. We leave the details as an exercise.

10. Although the recurrence makes it unnecessary to have an explicit representation of the Chebyshev polynomials to evaluate Chebyshev series, we will need their coefficients later when we turn to the economization of power series. They can be computed by a simple algorithm. To derive it, write

$$c_k(t) = c_{0k} + c_{1k}t + c_{2k}t^2 + \cdots + c_{kk}t^k.$$

The recurrence (4.5) can be written schematically in the form

$$
\begin{array}{llll}
-c_{k-2}(t) = & -c_{0,k-2} - & c_{1,k-2}t - & c_{2,k-2}t^2 - \cdots \\
+2tc_{k-1}(t) = & & 2c_{0,k-1}t + & 2c_{1,k-1}t^2 + \cdots \\
\hline
c_k(t) = & -c_{0,k-2} + (2c_{0,k-1} - c_{1,k-2})t + & (2c_{1,k-1} - c_{2,k-2})t^2 + \cdots
\end{array}
$$

From this we see that the general equation for the coefficients of $c_k$ are given by

$$c_{0k} = -c_{0,k-2}$$

and

$$c_{ik} = 2c_{i-1,k-1} - c_{i,k-2}, \qquad i = 1, \ldots, k.$$

The following program implements this recursion, accumulating the coefficients in an upper triangular array $C$.

1.   $C[0,0] = 1$
2.   $C[0,1] = 0$; $C[1,1] = 1$
3.  **for** $k = 2$ **to** $m$
4.     $C[0,k] = -C[0,k-2]$
5.     **for** $i = 1$ **to** $k$
6.       $C[i,k] = 2*C[i-1,k-1] - C[i,k-2]$
7.     **end for** $i$
8.  **end for** $k$

$$(4.6)$$

The algorithm requires $O(m^2)$. It is not as efficient as it could be, since it does not take advantage of the fact that every other coefficient of a Chebyshev polynomial is zero.

## Economization of power series

11. We now turn to obtaining a Chebyshev expansion of the function $f$ we wish to approximate. The following technique assumes that $f$ has a power series expansion:

$$f(t) = a_0 + a_1 t + a_2 t^2 + \cdots.$$

Suppose we want to approximate $f$ to accuracy $\epsilon$. We first truncate the series at a point where its error is well below $\epsilon$:

$$f(t) \cong a_0 + a_1 t + a_2 t^2 + \cdots + a_m t^m.$$

We then rewrite the series in terms of Chebyshev polynomials

$$a_0 + a_1 t + a_2 t^2 + \cdots + a_m t^m = b_0 c_0(t) + b_1 c_1(t) + b_2 c_2(t) + \cdots + b_m c_m(t) \quad (4.7)$$

(we will show how this can be done in a moment). Finally, we look at the $b_k$ (assuming they decrease reasonably rapidly), choose one that is sufficiently small, and truncate the Chebyshev sum to give our approximation

$$p(t) = b_0 c_0(t) + b_1 c_1(t) + b_2 c_2(t) + \cdots + b_k c_k(t).$$

12. We have already explained that if the $b_i$ decrease quickly enough, the term $b_{k+1} c_{k+1}(t)$ will dominate, and owing to the equi-alternation properties of $c_{k+1}$, we will obtain a nearly optimal approximation to the sum $a_0 + a_1 t + a_2 t^2 + \cdots + a_m t^m$. Moreover, since $\|c_{k+1}\|_\infty = 1$, the error in this approximation will be approximately $b_k$. Hence we should choose $k$ so that $|b_k|$ is a little less than $\epsilon$.

13. In the bad old days before computers, the conversion from the power sum to the Chebyshev sum in (4.7) would have been done by hand — a tedious and error-prone procedure. Since the work increases with $m$, the person doing the approximation would try to get away with the smallest possible value of $m$. None of this is necessary today. The conversion can be accomplished by matrix operations, and for practical purposes there is no need to restrict $m$.

14. First some notation. Let us denote by $C$ the $(m+1) \times (m+1)$ upper triangular matrix of coefficients generated by (4.6) for $k = m$. Let us also introduce the row vectors of functions

$$\mathbf{t}^{\mathrm{T}} = (1 \ t \ t^2 \ \cdots \ t^m) \quad \text{and} \quad \mathbf{c}^{\mathrm{T}} = (c_0(t) \ c_1(t) \ c_2(t) \ \cdots \ c_m(t)).$$

Then it is easily seen that

$$\mathbf{c}^{\mathrm{T}} = \mathbf{t}^{\mathrm{T}} C. \quad (4.8)$$

Moreover, if we set

$$a = \begin{pmatrix} a_0 \\ a_1 \\ \vdots \\ a_m \end{pmatrix} \quad \text{and} \quad b = \begin{pmatrix} b_0 \\ b_1 \\ \vdots \\ b_m \end{pmatrix},$$

then (4.7) can be written in the form

$$\mathbf{t}^{\mathrm{T}} a = \mathbf{c}^{\mathrm{T}} b.$$

From (4.8) we have $\mathbf{t}^{\mathrm{T}} = \mathbf{c}^{\mathrm{T}} C^{-1}$, and hence

$$\mathbf{c}^{\mathrm{T}} C^{-1} a = \mathbf{c}^{\mathrm{T}} b.$$

Since the components of $\mathbf{c}$ are independent functions of $t$, we have $C^{-1}a = b$. Hence we can compute $b$ by solving the triangular system

$$Cb = a.$$

This system requires $O(m^2)$ arithmetic operations for its solution — the same as for the generation of $C$. Such calculations are easily within the capabilities of the lowliest computer for $m = 100$. Consequently, we are effectively unlimited in the size of $m$.

15. Let us illustrate this procedure for the function $e^t$ on $[-1, 1]$. The following table gives data on the approximations obtained from a power series truncated at $m = 20$ — large enough so that the accuracy of the series itself is not in question.

| $k$ | $b_{k+1}$ | $\|p - f\|_\infty$ | V-P ratio |
|---|---|---|---|
| 1 | 2.7e−01 | 3.2e−01 | 1.39 |
| 2 | 4.4e−02 | 5.0e−02 | 1.28 |
| 3 | 5.4e−03 | 6.0e−03 | 1.22 |
| 4 | 5.4e−04 | 5.9e−04 | 1.18 |
| 5 | 4.4e−05 | 4.8e−05 | 1.15 |
| 6 | 3.1e−06 | 3.4e−06 | 1.13 |
| 7 | 1.9e−07 | 2.1e−07 | 1.12 |
| 8 | 1.1e−08 | 1.1e−08 | 1.11 |
| 9 | 5.5e−10 | 5.7e−10 | 1.10 |
| 10 | 2.4e−11 | 2.6e−11 | 1.09 |

You can see that the size of the coefficient $b_{k+1}$ is only a slight underestimate of the actual error. The V-P ratios are near one. They tell us, for example, that if we want to approximate $e^x$ to an accuracy of $10^{-5}$, we must take $k = 6$. For $k = 5$ the best approximation is no better than $4.8 \cdot 10^{-5}/1.15 = 4.2 \cdot 10^{-5}$.

## Farewell to $C[a, b]$

16. This lecture completes our tour of uniform approximation, but it by no means exhausts the subject. There are other ways of obtaining nearly alternating approximations. Two of them are least squares expansion in a Chebyshev series and interpolation at the zeros of the Chebyshev polynomials.

17. The chief use of uniform approximation is the evaluation of special functions. For these purposes, polynomials are often too restrictive. For example, our linear sine does not share with $\sin t$ the important property that

$\lim_{t \to 0} t^{-1} \sin t = 1$. One cure is to write

$$\sin t = t\left(1 - \frac{t^2}{6} + \frac{t^4}{120} - \cdots\right)$$

and economize the series in the parentheses. This does not give a best approximation — but it will give a better approximation.

18. You should also consider rational approximations, for which there is a well-developed theory. Rational functions can mimic the effects of nearby poles. As a trivial example, economization of power series is impossible on the function $1/(1 + x^2)$ in the interval $[-5, 5]$, since the radius of convergence of its power series is one. But the function itself is rational.

# Approximation

Discrete, Continuous, and Weighted Least Squares
Inner-Product Spaces
Quasi-Matrices
Positive Definite Matrices
The Cauchy and Triangle Inequalities
Orthogonality
The QR Factorization

## Discrete, continuous, and weighted least squares

1. In Lecture 1 we showed how discrete least squares approximation could be used to determine proportions of radioactive elements having known decay constants. More generally, the least squares problem in $\mathbf{R}^n$ has the following form. Given linearly independent vectors $x_1, \ldots, x_k$ in $\mathbf{R}^n$ and a vector $y$ also in $\mathbf{R}^n$ determine constants $\beta_1, \ldots, \beta_k$ such that

$$\|y - \beta_1 x_1 - \cdots - \beta_k x_k\|_2$$

is minimized. Here

$$\|u\|_2^2 = u^{\mathrm{T}} u = \sum_{i=1}^{n} u_i^2$$

is the usual 2-norm on $\mathbf{R}^n$.

2. There are continuous analogues of the discrete least squares problem. For example, suppose we want to approximate a function $y(t)$ by a polynomial of degree $k - 1$ on $[-1, 1]$. If we define

$$x_k(t) = t^{k-1}$$

and for any square-integrable function $u$ set

$$\|u\|_2^2 = \int_{-1}^{1} u(t)^2 \, dt,$$

then our problem once again reduces to minimizing $\|y - \beta_1 x_1 - \cdots - \beta_k x_k\|_2^2$.

3. In certain applications some parts of an approximation may be more important than others. For example, if the points $y_i$ in a discrete least squares problem vary in accuracy, we would not want to see the less accurate data contributing equally to the final fit. We can handle this problem by redefining the norm. Specifically, let $w_i$ be a *weight* that reflects the accuracy of $y_i$. (It

might, for example, be proportional to the inverse of the standard deviation of the error in $y_i$.) If we define

$$\|u\|^2 = \sum_{i=1}^{n} w_i^2 u_i^2,$$

then the inaccurate $y$'s will have lessened influence in the minimization of $\|y - \beta_1 x_1 - \cdots - \beta_k x_k\|$. As an extreme example, if $w_i = 0$, then $y_i$ has no influence at all.

Continuous problems can also be weighted. For example, if we desire to emphasize the endpoints of the interval $[-1, 1]$ we might define a norm by

$$\|u\|^2 = \int_{-1}^{1} \frac{u(t)^2}{\sqrt{1 - t^2}} \, dt.$$

(To see that this improper integral is well defined, make the substitution $t = \cos s$.) Since the weight function $1/\sqrt{1 - t^2}$ asymptotes to infinity at $-1$ and $1$, it will force a better fit around $-1$ and $1$ than around zero.

## Inner-product spaces

4. The norms defined above turn their vector spaces into normed linear spaces. But they are normed linear spaces with a special structure; for in all cases the norm can be generated by a more general function of two vectors called an *inner product*. For the four problems mentioned above, these functions are

    1.  $(x, y) = x^{\mathrm{T}} y = \sum_i x_i y_i,$

    2.  $(x, y) = \int_{-1}^{1} x(t) y(t) \, dt,$

    3.  $(x, y) = \sum_i w_i^2 x_i y_i,$

    4.  $(x, y) = \int_{-1}^{1} x(t) y(t) (1 - t^2)^{-\frac{1}{2}} \, dt.$

In all four cases the norm is given by

$$\|x\|^2 = (x, x).$$

5. Abstracting from these four examples, we now define an *inner-product space* as a vector space $\mathcal{V}$ and an inner product $(\cdot, \cdot) : \mathcal{V} \times \mathcal{V} \to \mathbf{R}$. The inner product has the following properties.

    1. Symmetry:    $(x, y) = (y, x).$

    2. Linearity:    $(x, \alpha y + \beta z) = \alpha(x, y) + \beta(x, z).$

    3. Definiteness:  $x \neq 0 \implies (x, x) > 0.$

Symmetry and linearity imply that the inner product is linear in its first argument as well as its second.[6]

6. Given an inner-product space, we can define a norm on it by $\|x\|^2 = (x,x)$.

To qualify as a norm, the function $\|\cdot\|$ must be definite and homogeneous and must satisfy the triangle inequality (see §1.17). The fact that $\|\cdot\|$ is definite and homogeneous follows immediately from the definition of the inner product. The triangle inequality requires more work. However, before we derive it, we must dispose of some technical problems.

## Quasi-matrices

7. The elementary theory of approximation in inner-product spaces is essentially the same for discrete and continuous spaces. Yet if you compare a book on numerical linear algebra with a book on approximation theory, you will find two seemingly dissimilar approaches. The numerical linear algebra book will treat the material in terms of matrices, whereas the approximation theory book will be cluttered with scalar formulas involving inner products — formulas that on inspection will be found to be saying the same thing as their more succinct matrix counterparts. Since the approaches are mathematically equivalent, it should be possible to find a common ground. We will do so first by introducing a notational change and second by extending the notion of matrix.

8. Part of the problem is that the conventional notation for the inner product is not compatible with matrix conventions. We will therefore begin by writing

$$x^{\mathrm{T}}y$$

for the inner product in all vector spaces — discrete or continuous. Note that this is an operational definition. It does not imply that there is an object $x^{\mathrm{T}}$ that multiplies $y$ — although it will be convenient at times to pretend there is such an object.

9. Another problem is that matrices, as they are usually understood, cannot be fashioned from vectors in a continuous space. In $\mathbf{R}^n$ we can build a matrix by lining up a sequence of vectors. For example, if $x_1, \ldots, x_k \in \mathbf{R}^n$, then

$$X = (x_1 \ x_2 \ \cdots \ x_k) \tag{5.1}$$

is a matrix with $n$ rows and $k$ columns. If we attempt to do the same thing with members of a continuous vector space we end up with an object that has no rows. For the moment let's call these objects quasi-matrices and denote the set of quasi-matrices with $k$ columns by $\mathcal{V}^k$. We have already seen quasi-matrices in §4.14, where we treated them as row vectors of functions. This works in a

---

[6]When the vector spaces are over the field of complex numbers, symmetry is replaced by conjugate symmetry: $(x,y) = \overline{(y,x)}$. More on this when we consider eigenvalue problems.

general normed linear space, but our range of operations with such objects is limited. In an inner-product space, on the other hand, quasi-matrices can be made to behave more like ordinary matrices.

10. The key to performing operations with quasi-matrices is to ask if the operation in question makes sense. For example, if $X \in \mathcal{V}^k$ is partitioned as in (5.1) and $b \in \mathbf{R}^k$, then the product

$$Xb = \beta_1 x_1 + \cdots + \beta_k x_k$$

is well defined and is a member of $\mathcal{V}$.

Similarly if $A = (a_1 \; a_2 \; \cdots \; a_\ell) \in \mathbf{R}^{k \times \ell}$, then the product

$$XA = (Xa_1 \; Xa_2 \; \cdots \; Xa_\ell)$$

is a well-defined quasi-matrix. On the other hand, the product $AX$ does not exist, since $\mathcal{V}$ does not have a matrix-vector product.

Two operations involving the inner product will be particularly important. If $X$ and $Y$ are quasi-matrices with $k$ and $\ell$ columns, then the product $X^{\mathrm{T}}Y$ is a $k \times l$ matrix defined by

$$X^{\mathrm{T}}Y = \begin{pmatrix} x_1^{\mathrm{T}}y_1 & x_1^{\mathrm{T}}y_2 & \cdots & x_1^{\mathrm{T}}y_\ell \\ x_2^{\mathrm{T}}y_1 & x_2^{\mathrm{T}}y_2 & \cdots & x_2^{\mathrm{T}}y_\ell \\ \vdots & \vdots & & \vdots \\ x_k^{\mathrm{T}}y_1 & x_k^{\mathrm{T}}y_2 & \cdots & x_k^{\mathrm{T}}y_\ell \end{pmatrix}.$$

If $X, Y \in \mathcal{V}^k$ then the product $XY^{\mathrm{T}}$ is an operator mapping $\mathcal{V}$ into $\mathcal{V}$ defined by

$$(XY^{\mathrm{T}})z = X(Y^{\mathrm{T}}z).$$

We will encounter this operator when we introduce projections.

11. The term "quasi-matrix" has served its purpose, and we will now give it an honorable burial. Its memorial will be the notation $\mathcal{V}^k$ to denote a matrix formed as in (5.1) from $k$ elements of the vector space $\mathcal{V}$.[7]

## Positive definite matrices

12. Positive definite matrices play an important role in approximation in inner-product spaces. Here we collect some facts we will need later.

---

[7]In the original lecture I gave a more formal development and even wrote quasi-matrices in small caps to distinguish them from ordinary matrices. This is bad business for two reasons. First, it represents a triumph of notation over common sense. Second, if you introduce a formal mathematical object, people are likely to start writing papers about it. Like "On the generalized inverse of quasi-matrices." Ugh!

A symmetric matrix $A$ is *positive semidefinite* if

$$x \neq 0 \implies x^{\mathrm{T}} A x \geq 0. \tag{5.2}$$

If strict equality holds in (5.2) for all nonzero $x$, then $A$ is *positive definite*.

13. The following three facts will be used in the sequel. The first two are easy to establish, as is the third for $2 \times 2$ matrices — the only case we will use.

> 1. A positive definite matrix is nonsingular.
> 2. Let $X \in \mathcal{V}^k$. Then $X^{\mathrm{T}} X$ is positive semidefinite. It is positive definite if and only if the columns of $A$ are linearly independent.
> 3. The determinant of a positive semidefinite matrix is nonnegative. The determinant of a positive definite matrix is positive.

## The Cauchy and triangle inequalities

14. We will now show that a norm defined by an inner product satisfies the triangle inequality. We begin with an inequality, called the Cauchy inequality, that is important in its own right.

> For any vector $x$ and $y$
> $$|x^{\mathrm{T}} y| \leq \|x\|\|y\| \tag{5.3}$$
> with equality if and only if $x$ and $y$ are linearly dependent.

To establish this result suppose that $x$ and $y$ are linearly independent. Then the matrix

$$A = (x \ y)^{\mathrm{T}} (x \ y) = \begin{pmatrix} x^{\mathrm{T}} x & x^{\mathrm{T}} y \\ y^{\mathrm{T}} x & y^{\mathrm{T}} y \end{pmatrix}$$

is positive definite. Hence

$$0 < \det(A) = (x^{\mathrm{T}} x)(y^{\mathrm{T}} y) - (x^{\mathrm{T}} y)^2 = (\|x\|^2 \|y\|^2) - (x^{\mathrm{T}} y)^2,$$

and (5.3) follows immediately.

On the other hand if vectors $x$ and $y$ are dependent, then $A$ is positive semidefinite, in which case $\det(A) = 0$, and equality holds in (5.3).

15. With the Cauchy inequality established, the triangle inequality is a foregone conclusion:

$$\begin{aligned} \|x + y\|^2 &= (x + y)^{\mathrm{T}} (x + y) \\ &= x^{\mathrm{T}} x + 2 x^{\mathrm{T}} y + y^{\mathrm{T}} y \\ &\leq \|x\|^2 + 2\|x\|\|y\| + \|y\|^2 \qquad \text{by the Cauchy inequality} \\ &= (\|x\|^2 + \|y\|^2). \end{aligned}$$

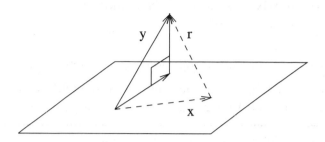

Figure 5.1. *Geometric illustration of least squares.*

## Orthogonality

16.  Orthogonality between vectors is a generalization of perpendicularity in Euclidean two- and three-dimensional space. The concept is central to approximation in inner-product spaces. To see why consider the problem of approximating a vector $y \in \mathbf{R}^3$ by a linear combination $\beta_1 x_1 + \beta_2 x_2$ of two linearly independent vectors $x_1$ and $x_2$. Our criterion for best approximation requires that we minimize the norm of the residual $r = y - \beta_1 x_1 - \beta_2 x_2$.

The geometric situation is illustrated in Figure 5.1. As $\beta_1$ and $\beta_2$ vary, the vector $x = \beta_1 x_1 + \beta_2 x_2$ varies in a two-dimensional subspace represented by the plane in the figure. The dashed lines show a typical approximation $x$ and the corresponding residual $r$. Geometric intuition tells us that the size of $r$ should be minimized when $r$ is at right angles to the plane, as illustrated by the solid lines.

17.  To generalize this insight, we must decide what we mean for a vector to be perpendicular to a subspace. We will begin by using the Cauchy inequality to define the angle between two vectors.

Since $\cos\theta$ ranges from 1 to $-1$ as $\theta$ ranges from 0 to $\pi$, the Cauchy inequality can be rewritten in the form

$$x^{\mathrm{T}}y = \cos\theta \|x\| \|y\|$$

for some unique $\theta \in [0, \pi]$. In ordinary Euclidean two- or three-dimensional space, it is easy to verify that $\theta$ is actually the angle between $x$ and $y$. In a general inner-product space we *define* $\theta$ to be the angle between $x$ and $y$.

18.  If $\theta = \frac{\pi}{2}$ — i.e., if $x$ and $y$ are at right angles — then $\cos\theta$ and hence $x^{\mathrm{T}}y$ are zero. Thus we will say that the vectors $x$ and $y$ are *orthogonal* $x^{\mathrm{T}}y = 0$. We write $x \perp y$.

Note that we do not require $x$ and $y$ to be nonzero in our definition of orthogonality. It is useful to have the zero vector orthogonal to all vectors, and even more useful to have the only self-orthogonal vector be the zero vector.

19. A vector $q$ is *normalized* if $\|q\| = 1$. Any nonzero vector can be normalized by dividing it by its norm.

The vectors $q_1, q_2, \ldots, q_k$ are said to be *orthonormal* if they are normalized and are pairwise orthogonal. In other words,

$$q_i^{\mathrm{T}} q_j = \begin{cases} 1 & \text{if } i = j, \\ 0 & \text{if } i \neq j. \end{cases} \tag{5.4}$$

If we set $Q = (q_1 \; q_2 \; \cdots \; q_k)$, then (5.4) may be written succinctly in the form

$$Q^{\mathrm{T}} Q = I_k,$$

where $I_k$ is the identity matrix of order $k$.

If $X \in \mathcal{V}^k$ then $q$ is orthogonal to the columns of $X$ if and only if $X^{\mathrm{T}} q = 0$.

20. A fact that goes deeper that these essentially notational results is the *Pythagorean equality*. It is the analytic equivalent of the Pythagorean theorem, which says that the square of the hypotenuse of a right triangle is equal to the sum of its squares of its two sides. Analytically expressed:

> If $x$ and $y$ are orthogonal then
> $$\|x + y\|^2 = \|x\|^2 + \|y\|^2. \tag{5.5}$$

The proof is easy:

$$\begin{aligned}
\|x + y\|^2 &= (x + y)^{\mathrm{T}}(x + y) \\
&= x^{\mathrm{T}} x + 2 x^{\mathrm{T}} y + y^{\mathrm{T}} y \\
&= x^{\mathrm{T}} x + y^{\mathrm{T}} y \qquad \text{since } x \perp y \\
&= \|x\|^2 + \|y\|^2.
\end{aligned}$$

## The QR factorization

21. Every finite-dimensional subspace $\mathcal{X} \subset \mathcal{V}$ has a basis — say, $x_1, \ldots, x_k$. The *QR factorization* shows that this basis can be converted into an orthonormal basis for $\mathcal{X}$. Specifically:

> Let $X \in \mathcal{V}^k$ have linearly independent columns. Then there is a $Q \in \mathcal{V}^k$ with orthonormal columns and an upper triangular matrix $R$ of order $k$ having positive diagonal elements such that
> $$X = QR.$$
> Both $Q$ and $R$ are unique.

22.  Before establishing the existence of the QR factorization, let's see what it means. Since $R$ is nonsingular, we can write $XS = Q$, where $S = R^{-1}$. In other words,

$$
(q_1 \ q_2 \ \cdots \ q_k) = (x_1 \ x_2 \ \cdots \ x_k) \begin{pmatrix} s_{11} & s_{12} & \cdots & s_{1k} \\ 0 & s_{22} & \cdots & s_{2k} \\ \vdots & \vdots & & \vdots \\ 0 & 0 & \cdots & s_{kk} \end{pmatrix},
$$

or equivalently

$$
\begin{aligned}
q_1 &= s_{11}x_1, \\
q_2 &= s_{12}x_1 + s_{22}x_2, \\
&\quad\cdot\quad\cdot\quad\cdot \\
q_k &= s_{1k}x_1 + s_{2k}x_2 + \cdots + s_{kk}x_k.
\end{aligned}
$$

Thus $q_1$ is just $x_1$ normalized. The vector $q_2$ is a linear combination of $x_1$ and $x_2$ that is orthogonal to $q_1$ or equivalently to $x_1$. In general, $q_j$ is a linear combination of $x_1, \ldots, x_j$ that is orthogonal to $q_1, \ldots, q_{j-1}$, or equivalently to $x_1, \ldots, x_{j-1}$. Thus the QR factorization amounts to a sequential orthogonalization of the vectors $x_1, \ldots, x_k$. The term sequential is important here. Having orthogonalized $x_1, \ldots, x_k$ we have also orthogonalized $x_1, \ldots, x_l$ for any $l < k$.

# Approximation

Existence and Uniqueness of the QR Factorization
The Gram–Schmidt Algorithm
Projections
Best Approximation on Inner-Product Spaces

## Existence and uniqueness of the QR factorization

1. The proof of the existence of the QR factorization proceeds by induction on $k$. For $k = 1$, $Q$ is just the vector $x_1$ normalized and $R = (\|x_1\|)$.

Now assume that the theorem is true for $k - 1 \geq 1$. We seek a QR factorization of $X = (x_1 \cdots x_k)$ in the partitioned form

$$(X_1 \ x_k) = (Q_1 \ q_k) \begin{pmatrix} R_{11} & r_{1k} \\ 0 & \rho_{kk} \end{pmatrix}.$$

(The indices of a submatrix in the partition are the same as the element in the upper left of the submatrix. This scheme is called *northwest indexing,* and, as we shall see, the scheme is very useful in keeping track of index ranges while coding matrix algorithms.) On computing the first column of this partition, we find that

$$X_1 = Q_1 R_{11}.$$

This is the QR factorization of $X_1$, which exists by the induction hypothesis.

The second column of the partition is

$$x_k = Q_1 r_{1k} + \rho_{kk} q_k.$$

Multiplying this equation by $Q_1^{\mathrm{T}}$ we find that

$$Q_1^{\mathrm{T}} x_k = r_{1k} + \rho_{kk} Q_1^{\mathrm{T}} q_k.$$

From this we see that $\rho_{kk} q_k$ will be orthogonal to the columns of $Q_1$ if and only if

$$r_{1k} = Q_1^{\mathrm{T}} x_k, \tag{6.1}$$

in which case

$$\rho_{kk} q_k = x_k - Q_1 r_{1k}. \tag{6.2}$$

If $x_k - Q_1 r_{1k} \neq 0$, we can take

$$\rho_{kk} = \|x_k - Q_1 r_{1k}\|, \tag{6.3}$$

39

which normalizes $q_k$. But since $Q_1 = X_1 R_{11}^{-1}$,

$$x_k - Q_1 r_{1k} = x_k - X_1(R_{11}^{-1} r_{1k}).$$

The side of this equation is a nontrivial linear combination of the independent vectors $x_1, \ldots, x_k$ and hence is nonzero.

The uniqueness of the factorization follows from the uniqueness of the factorization $X_1 = Q_1 R_1$ and the fact that (6.1), (6.3), and (6.2) uniquely determine $r_{1k}$, $\rho_{kk}$, and $q_k$.

## The Gram–Schmidt algorithm

2.     The proof of the existence of the QR decomposition is constructive, and the resulting algorithm is called the Gram–Schmidt algorithm. To derive it let us suppose that we have computed the first $j-1$ columns of $Q$ — which we will denote by $Q[:, 1:j-1]$ — and the leading principal minor $R[1:j-1, 1:j-1]$ of $R$. Our problem then is to compute the $j$th columns of $Q$ and $R$. From (6.1) we see that

$$R[1:j-1, j] = Q[:, 1:j-1]^{\mathrm{T}} * X[:, j].$$

From (6.2) we see that

$$Q[:, j] = (X[:, j] - Q[:, 1:j-1] * R[1:j-1, j])/R[j, j],$$

where from (6.3)

$$R[j, j] = \|X[:, j] - Q[:, 1:j-1] * R[1:j-1, j]\|.$$

We summarize this algorithm in the following program.

1.   **for** $j = 1$ **to** $k$
2.       $Q[:, j] = X[:, j]$
3.       $R[1:j-1, j] = Q[:, 1:j-1]^{\mathrm{T}} * Q[:, j]$
4.       $Q[:, j] = Q[:, j] - Q[:, 1:j-1] * R[1:j-1, j]$
5.       $R[j, j] = \|Q[:, j]\|$
6.       $Q[:, j] = Q[:, j]/R[j, j]$
7.   **end for** $j$

3.   An important variant of the Gram–Schmidt algorithm can be obtained by expanding statements 3 and 4 in loops.

1.   **for** $j = 1$ **to** $k$
2.      $Q[:,j] = X[:,j]$
3.      **for** $i = 1$ **to** $j-1$
4.         $R[i,j] = Q[:,i]^{\mathrm{T}}*Q[:,j]$
5.      **end for** $i$
6.      **for** $i = 1$ **to** $j-1$            (6.4)
7.         $Q[:,j] = Q[:,j] - Q[:,i]*R[i,j]$
8.      **end for** $i$
9.      $R[j,j] = \|Q[:,j]\|$
10.     $Q[:,j] = Q[:,j]/R[j,j]$
11.  **end for** $j$

It is a remarkable fact that if we combine the two inner loops, we compute the same QR factorization.

1.   **for** $j = 1$ **to** $k$
2.      $Q[:,j] = X[:,j]$
3.      **for** $i = 1$ **to** $j-1$
4.         $R[i,j] = Q[:,i]^{\mathrm{T}}*Q[:,j]$
5.         $Q[:,j] = Q[:,j] - Q[:,i]*R[i,j]$
6.      **end for** $i$
7.      $R[j,j] = \|Q[:,j]\|$
8.      $Q[:,j] = Q[:,j]/R[j,j]$
9.   **end for** $j$

This algorithm is called the *modified Gram–Schmidt algorithm*. Note that it is really a different algorithm, since $Q[:,j]$ changes as we compute the successive values of $R[i,j]$, whereas in (6.4) it remains constant until all the $R[i,j]$ have been computed. We leave it as a (mandatory) exercise to verify that it works.

4. The Gram–Schmidt algorithm can be used in both continuous and discrete spaces. Unfortunately, in the discrete case it is numerically unstable and can give vectors that are far from orthogonal. The modified version is better, but it too can produce nonorthogonal vectors. We will later give an algorithm for the discrete case that preserves orthogonality.

## Projections

5. Figure 5.1 suggests that the best approximation in a subspace to a vector $y$ will be the shadow cast by $y$ at high noon on the subspace. Such shadows are called projections. We are going to show how to use the QR factorization to compute projections.

6. Let the columns of $X \in \mathcal{V}^k$ be linearly independent, and let $X = QR$ be the QR factorization of $X$. Then the columns of $Q$ form a basis for the space $\mathcal{X}$ spanned by the columns of $X$. Let

$$P_X = QQ^{\mathrm{T}} \quad \text{and} \quad P_\perp = I - P_X. \tag{6.5}$$

The following result shows that these two matrices are related to the geometry of the space $\mathcal{X}$.

---

Let $P_X$ and $P_\perp$ be defined by (6.5). Then

  1. $x \in \mathcal{X} \implies P_X x = x$ and $P_\perp x = 0$.
  2. $z \perp \mathcal{X} \implies P_X z = 0$ and $P_\perp z = z$.

---

To prove the first of the above results, note that if $x \in \mathcal{X}$ then $x = Qb$ for some $b$ (because the columns of $Q$ span $\mathcal{X}$). Hence

$$P_X x = (QQ^{\mathrm{T}})(Qb) = Q(Q^{\mathrm{T}}Q)b = Qb = x.$$

Moreover, if $P_X x = x$, then $P_\perp x = (I - P_X)x = x - x = 0$. The second result is established similarly.

7. Let $y$ be an arbitrary vector in $\mathcal{V}$ and let

$$y_X = P_X y \quad \text{and} \quad y_\perp = P_\perp y.$$

Then since $P_X + P_\perp = I$,

$$y = y_X + y_\perp. \tag{6.6}$$

Now $y_X = Q(Q^{\mathrm{T}}y)$ is a linear combination of the columns of $Q$ and hence lies in $\mathcal{X}$. Moreover,

$$Q^{\mathrm{T}}y_\perp = Q^{\mathrm{T}}(I - QQ^{\mathrm{T}})y = Q^{\mathrm{T}}y - (Q^{\mathrm{T}}Q)Q^{\mathrm{T}}y = Q^{\mathrm{T}}y - Q^{\mathrm{T}}y = 0.$$

Hence the vector $y_\perp$ is orthogonal to $\mathcal{X}$. Thus (6.6) decomposes the vector $y$ into the sum of two orthogonal vectors, one lying in $\mathcal{X}$ and the other orthogonal to $\mathcal{X}$. (This, in fact, is just the decomposition illustrated in Figure 5.1.)

Such a decomposition is unique. For if $y = \hat{y}_X + \hat{y}_\perp$ is another such decomposition, we have from the result of §6.6

$$y_X = P_X y = P_X(\hat{y}_X + \hat{y}_\perp) = \hat{y}_X,$$

since $\hat{y}_X \in \mathcal{X}$ and $\hat{y}_\perp \perp \mathcal{X}$.

To summarize:

---

For any vector $y$ the sum

$$y = P_X y + P_\perp y$$

decomposes $y$ uniquely into the sum of a vector lying in $\mathcal{X}$ and a vector orthogonal to $\mathcal{X}$.

---

8. The vector $y_X = P_X y$ is called *the orthogonal projection of $y$ onto the column space of $X$*, and the vector $y_\perp = P_\perp y$ is called *complementary projection*. The operators that produce these vectors are called *projection operators* or *projectors*. They are important in many numerical applications. Here we shall use them to solve the best approximation problem.

## Best approximation on inner-product spaces

9. We now are in a position to give a complete solution of the approximation problem in an inner-product space. Here are the ingredients.

1. A vector space $\mathcal{V}$ and an inner product $(x, y)$. As usual, we shall write $x^T y$ for $(x, y)$.

2. A vector $y$ to be approximated.

3. Linearly independent vectors $x_1, \ldots, x_k$ to do the approximating.

We look for an approximation in the form

$$y \cong \beta_1 x_1 + \beta_2 x_2 + \cdots + \beta_k x_k.$$

The best approximation will minimize $\|y - \beta_1 x_1 - \beta_2 x_2 - \cdots - \beta_k x_k\|$, where $\|x\|^2 = x^T x$.

10. To bring matrix techniques into play, let

$$X = (x_1 \ x_2 \ \cdots \ x_k) \quad \text{and} \quad b = (\beta_1 \ \beta_2 \ \cdots \ \beta_k)^T.$$

Then we wish to determine $b$ to

$$\text{minimize } \|y - Xb\|.$$

11. The solution of the problem involves the use of projections and the Pythagorean equality (5.5):

$$\begin{aligned}
\|y - Xb\|^2 &= \|P_X(y - Xb) + P_\perp(y - Xb)\|^2 && (P_X + P_\perp = I) \\
&= \|P_X(y - Xb)\|^2 + \|P_\perp(y - Xb)\|^2 && (\text{Pythagorean equality}) \\
&= \|P_X y - Xb\|^2 + \|P_\perp y\|^2 && (P_X X = X, P_\perp X = 0).
\end{aligned}$$

Now the second term in the last expression is independent of $b$. Consequently, we can minimize $\|y - Xb\|$ by minimizing $\|P_X y - Xb\|$. But $P_X y$ is in the space spanned by the columns of $X$. Since the columns of $X$ are linearly independent, there is a unique vector $b$ such that

$$Xb = P_X y. \tag{6.7}$$

For this $b$, we have $\|P_X y - Xb\| = 0$, which is as small as a norm can get. Consequently, the vector $b$ satisfying (6.7) is the unique solution of our approximation problem.

12. Since our object is to derive numerical algorithms, we must show how to solve (6.7). There are two ways — one classical (due to Gauss and Legendre) and one modern.

Let $\mathcal{V}$ be an inner-product space. Let $X \in \mathcal{V}^n$ have linearly independent columns, and let $X = QR$ be the QR factorization of $X$.
Let $y \in \mathcal{V}$ and consider the problem

$$\text{minimize } \|y - Xb\|. \tag{6.8}$$

1.  The problem (6.8) has a unique solution $b$.
2.  The vector $b$ satisfies the normal equations

$$(X^{\mathrm{T}}X)b = X^{\mathrm{T}}y.$$

3.  The vector $b$ satisfies the QR equation

$$Rb = Q^{\mathrm{T}}y. \tag{6.9}$$

4.  The approximation $Xb$ is the projection of $y$ onto the column
    space of $X$.
5.  The residual vector $y - Xb$ is orthogonal to the column space of
    $X$.

Figure 6.1. *Summary of best approximation in an inner-product space.*

13. To derive the classical way, note that $X^{\mathrm{T}}P_X = X^{\mathrm{T}}$. Hence on multiplying
(6.7) by $X^{\mathrm{T}}$ we obtain

$$(X^{\mathrm{T}}X)b = X^{\mathrm{T}}y.$$

This is a $k \times k$ system of linear equations for $b$. They are called the *normal
equations*.

It is worth noting that the normal equations are really a statement that
the residual $y - Xb$ must be orthogonal to the column space of $X$, which is
equivalent to saying that $X^{\mathrm{T}}(y - Xb) = 0$.

Since the columns of $X$ are linearly independent, the matrix $X^{\mathrm{T}}X$ is
positive definite. Consequently, the normal equations can be solved by the
Cholesky algorithm.

14. The modern method is based on the QR factorization $X = QR$. Since the
columns of $Q$ are in the column space of $X$, we have $Q^{\mathrm{T}}P_X = Q^{\mathrm{T}}$. Moreover,
$Q^{\mathrm{T}}X = Q^{\mathrm{T}}QR = R$. Hence on multiplying (6.7) by $Q^{\mathrm{T}}$ we get

$$Rb = Q^{\mathrm{T}}y.$$

This system of equations is called the *QR equation*.

Since $R$ is upper triangular, the solution of the QR equation is trivial. It
might seem a disadvantage that we must first compute a QR factorization of

$X$. However, to form the normal equations we must compute the *cross-product matrix* $X^{\mathrm{T}}X$, which usually involves the same order of labor. We will return to a detailed comparison of the methods when we consider the discrete case.

15. A summary of our results on best approximations in an inner-product space is given in Figure 6.1.

# Approximation

Expansions in Orthogonal Functions
Orthogonal Polynomials
Discrete Least Squares and the QR Decomposition

## Expansions in orthogonal functions

1. At this point continuous and discrete approximation problems part ways. The reason is the nature of their inner products. The inner product in a discrete problem is a sum of products of scalars — something that can be readily computed. Consequently, once we have formed the matrices that define a problem, we can forget where they came from and proceed with a general algorithm.

The inner products for continuous problems, on the other hand, are integrals, which in principle must be treated analytically. Depending on the functions involved, this may be easy or difficult; but it involves more intellectual investment than computing a discrete inner product. Consequently, in solving continuous problems a good deal of effort is devoted to minimizing the number of inner products to be evaluated. The chief technique for doing this is to write the solution in terms of orthogonal functions.

2. The theoretical basis for the technique is the observation that if the columns of $X$ are orthonormal, then $X^{\mathrm{T}}X = I$. Hence from the normal equations we have

$$b = X^{\mathrm{T}}y.$$

Consequently, $\beta_i = x_i^{\mathrm{T}}y$; i.e., the problem can be solved by computing $k$ inner products. In this context, the coefficients $\beta_i$ are called *generalized Fourier coefficients*.

It is a matter of consequence that the coefficients $\beta_i$ are determined independently of one another. This means that having computed an approximation in terms of orthogonal functions, we can add terms at only the additional cost of computing the new coefficients.

3. For a specific example, let us consider the problem of expanding a function $f$ on $[-\pi, \pi]$ in the *Fourier series*

$$\begin{aligned} f(t) = a_0 + a_1 \cos t + a_2 \cos 2t + \cdots \\ b_1 \sin t + b_2 \sin 2t + \cdots. \end{aligned} \tag{7.1}$$

The inner product is the usual one that generates the continuous 2-norm:

$$(f, g) = \int_{-\pi}^{\pi} f(t)g(t)\, dt.$$

The first thing is to verify that the cosines and sines are indeed orthogonal. Consider the equations

$$\cos(i+j)t = \cos it \cos jt - \sin it \sin jt,$$
$$\cos(i-j)t = \cos it \cos jt + \sin it \sin jt. \tag{7.2}$$

Adding these two equations, we get

$$\cos(i+j)t + \cos(i-j)t = 2\cos it \cos jt.$$

From this it is easily seen that

$$\int_{-\pi}^{\pi} \cos it \cos jt \, dt = \begin{cases} \pi & \text{if } i = j > 0, \\ 0 & \text{if } i \neq j. \end{cases}$$

This establishes the orthogonality of cosines with themselves and shows that $\|\cos it\|^2 = \pi$ for $i > 0$. (The case $i = 0$ is an exception: $\|\cos 0t\|^2 = 2\pi$.)

The orthogonality of sines may be obtained by subtracting the equations (7.2) and integrating. We also find that $\|\sin it\|^2 = \pi$. The orthogonality between sines and cosines may be derived similarly from the identity

$$\sin(i+j)t = \sin it \cos jt + \cos it \sin jt.$$

4. Thus the Fourier coefficients in (7.1) are

$$a_0 = \frac{1}{2\pi} \int_{-\pi}^{\pi} f(t) \, dt,$$

and

$$a_i = \frac{1}{\pi} \int_{-\pi}^{\pi} f(t) \cos it \, dt, \quad b_i = \frac{1}{\pi} \int_{-\pi}^{\pi} f(t) \sin it \, dt, \qquad i > 0.$$

In some cases we may be able to evaluate these integrals analytically. Otherwise, we must use numerical integration.

5. Our general approach to approximation in inner-product spaces shows that when we truncate the Fourier series derived above we obtain the best normwise approximation by a linear combination of sines and cosines. However, the theory does not tell us how good the approximation is, or even whether it converges as we retain more and more terms. The study of the convergence of Fourier series is one of great theoretical, practical, and historical importance.

## Orthogonal polynomials

6. Orthogonal polynomials form another class of orthogonal functions. They not only are used to approximate functions, but they form the underpinnings of some of the best numerical algorithms. We cannot give a complete treatment

of orthogonal polynomials — even of the elementary properties. Instead we will give enough for the applications to follow.

7. Two polynomials $p$ and $q$ are orthogonal on the interval $[a, b]$ with respect to the weight function $w(t) > 0$ if

$$\int_a^b p(t)q(t)w(t)\, dt = 0.$$

The intervals may be improper. For example, either $a$ or $b$ or both may be infinite, in which case the weight function must decrease rapidly enough to make the resulting integrals meaningful.

8. We have already encountered a sequence of orthogonal polynomials: the Chebyshev polynomials $c_k(t)$, which satisfy the orthogonality conditions

$$\int_{-1}^1 \frac{c_i(t)c_j(t)}{\sqrt{1 - t^2}}\, dt = 0, \qquad i \neq j.$$

This fact can be easily verified from the trigonometric form (4.3) of the Chebyshev polynomials.

9. In §4.6 we showed that Chebyshev polynomials satisfy the simple recurrence relation

$$c_{k+1}(t) = 2tc_k(t) - c_{k-1}(t).$$

A recurrence like this allows one to generate values of a sequence of Chebyshev polynomials with the same order of work required to evaluate the very last one from its coefficients. This means that Chebyshev expansions of length $k$ can be evaluated in $O(k)$ time, as opposed to the $O(k^2)$ time that would be required to evaluate each polynomial in the expression.

Now the existence of this recurrence is not a special property of the Chebyshev polynomials but a consequence of their orthogonality. We are going to derive a general recurrence for an arbitrary sequence of orthogonal polynomials. The derivation is part of the formal theory of orthogonal polynomials. It does not depend on the interval of integration, the weight function, and certainly not on the variable of integration. Hence we will discard this baggage and write $\int pq$ for $\int_a^b p(t)q(t)w(t)\, dt = 0$.

10. We begin with two definitions. First, a sequence of polynomials $p_0, p_1, \ldots$ is said to be *basic* if $\deg(p_i) = i$ $(i = 0, 1, \ldots)$. Second, a basic sequence of polynomials is a *sequence of orthogonal polynomials* if its terms are pairwise orthogonal.

11. The existence of orthogonal polynomials is easy to establish. Just apply the Gram–Schmidt algorithm to the sequence $1, t, t^2, \ldots$. This also establishes the uniqueness up to normalization of the sequence of orthogonal polynomials, since the coefficients $r_{ij}$ in the Gram–Schmidt algorithms are uniquely

determined by the starting vectors and the requirement that the results be orthonormal.

12. We will need the following useful fact about basic sequences of polynomials.

> Let $p_0, p_1, \ldots$ be a basic sequence of polynomials, and let $q$ be a polynomial of degree $k$. Then $q$ can be written uniquely as a linear combination of $p_0, p_1, \ldots, p_k$.

We actually proved this result in §4.14, where we showed how to expand a truncated power series in Chebyshev polynomials. The $p_i$ correspond to the Chebyshev polynomials, and $q$ corresponds to the truncated power series. You should satisfy yourself that the result established in §4.14 depends only on the fact that the sequence of Chebyshev polynomials is basic.

13. A consequence of the above result is that:

> If $p_0, p_1, \ldots$ is a sequence of orthogonal polynomials then $p_i$ is orthogonal to any polynomial of degree less than $i$.

To see this, let $q$ be of degree $j < i$. Then $q$ can be written in the form $q = a_0 p_0 + \cdots + a_j p_j$. Hence

$$\int p_i q = \int p_i (a_0 p_0 + \cdots + a_j p_j) = a_0 \int p_i p_0 + \cdots + a_j \int p_i p_j = 0.$$

14. We are now in a position to establish the existence of the recurrence that generates the sequence of orthogonal polynomials. We will work with monic polynomials, i.e., polynomials whose leading coefficient is one.

> Let $p_0, p_1, p_2, \ldots$ be a sequence of monic orthogonal polynomials. Then there are constants $a_k$ and $b_k$ such that
>
> $$p_{k+1} = t p_k - a_k p_k - b_k p_{k-1}, \qquad k = 0, 1, \ldots.$$
>
> (For $k = 0$, we take $p_{-1} = 0$.)

The derivation of this *three-term recurrence* is based on the following two facts.

1. The orthogonal polynomials can be obtained by applying the Gram–Schmidt algorithm to any basic sequence of polynomials.[8]

2. We can defer the choice of the $k$th member of this basic sequence until it is time to generate $p_k$.

---

[8] Actually, we will use a variant of the algorithm in which the normalization makes the polynomials monic.

Specifically we are going to orthogonalize according to the following scheme.

Function to be orthogonalized   : $\quad 1 \qquad tp_0 \qquad tp_1 \qquad tp_2 \qquad \cdots$

$$\downarrow \quad \nearrow \quad \downarrow \quad \nearrow \quad \downarrow \quad \nearrow \quad \downarrow \quad \nearrow$$

Resulting orthogonal polynomial: $\quad p_0 \qquad p_1 \qquad p_2 \qquad p_3 \qquad \cdots$

As suggested above, we will not actually know $tp_k$ until we have generated $p_k$.

15. It is necessary (and instructive) to treat $p_0$ and $p_1$ separately. We will take $p_0 = 1$ and seek $p_1$ in the form

$$p_1 = tp_0 - a_0 p_0.$$

Since $p_1$ must be orthogonal to $p_0$, we must have

$$0 = \int p_0 (tp_0 - a_0 p_0) = \int tp_0^2 - a_0 \int p_0^2.$$

Hence

$$a_0 = \frac{\int tp_0^2}{\int p_0^2}.$$

The denominator in this expression is nonzero because $p_0 \neq 0$.

16. For the general case, we seek $p_{k+1}$ in the form

$$p_{k+1} = tp_k - a_k p_k - b_k p_{k-1} - c_k p_{k-2} - \cdots.$$

Since $p_{k+1}$ must be orthogonal to $p_k$, we have

$$0 = \int p_k p_{k+1} = \int tp_k^2 - a_k \int p_k^2 - b_k \int p_k p_{k-1} - c_k \int p_k p_{k-2} - \cdots.$$

By orthogonality $0 = \int p_k p_{k-1} = \int p_k p_{k-2} = \cdots$. Hence $\int tp_k^2 - a_k \int p_k^2 = 0$, or

$$a_k = \frac{\int tp_k^2}{\int p_k^2}. \qquad (7.3)$$

Since $p_{k+1}$ must be orthogonal to $p_{k-1}$, we have

$$0 = \int p_{k-1} p_{k+1} = \int tp_{k-1} p_k - a_k \int p_{k-1} p_k - b_k \int p_{k-1}^2 - c_k \int p_{k-1} p_{k-2} - \cdots.$$

Since $0 = \int p_{k-1} p_k = \int p_{k-1} p_{k-2} = \cdots$, we have

$$b_k = \frac{\int tp_{k-1} p_k}{\int p_{k-1}^2}. \qquad (7.4)$$

The same argument shows that

$$c_k = \frac{\int tp_{k-2} p_k}{\int p_{k-2}^2}.$$

But $tp_{k-2}$ is a polynomial of degree $k - 1$ and hence is orthogonal to $p_k$. It follows that $c_k = 0$. Similarly, all the coefficients down to the multiplier of $p_0$ are zero. What is left is the three-term recurrence

$$p_{k+1} = tp_k - a_k p_k - b_k p_{k-1}, \qquad k = 0, 1, \ldots,$$

where $a_k$ and $b_k$ are defined by (7.3) and (7.4).

17. This completes the derivation of the three-term recurrence. Its formal nature should now be obvious. All that is needed is a space with an inner product $(\cdot, \cdot)$ (here generated by $\int$), a linear operator $A$ that satisfies $(x, Ay) = (Ay, x)$ (here multiplication by $t$), and an element of that space to get things started (here $p_0$). Later we will encounter the three-term recurrence in an entirely different setting when we treat the Lanczos and conjugate gradient algorithms.

## Discrete least squares and the QR decomposition

18. When the underlying vector space is $\mathbf{R}^n$, our approximation problem becomes one of minimizing $\|y - Xb\|$, where $X$ is now an $n \times k$ matrix. Here we will be concerned with minimizing the 2-norm defined by

$$\|z\|_2^2 = \sum_i z_i^2.$$

This problem is called the least squares problem because one minimizes the sum of squares of the components of the residual $r = y - Xb$ (see §1.5).

19. The fact that $X$ is a matrix — not a quasi-matrix — means that we have the additional option of premultiplying $X$ by a matrix — or equivalently of performing row operations on $X$. This flexibility means that there are more algorithms for solving discrete least squares problems than for solving continuous problems. We are going to describe an algorithm based on a generalization of the QR factorization called the QR decomposition. But first a definition.

20. A square matrix $Q$ is *orthogonal* if

$$Q^{\mathrm{T}} Q = I.$$

In other words, $Q$ is orthogonal if its inverse is its transpose. Since the left and right inverses of a matrix are the same, it follows that

$$QQ^{\mathrm{T}} = I.$$

Thus, an orthogonal matrix, in addition to having orthonormal columns, has orthonormal rows.

An important property of orthogonal matrices is that they preserve the 2-norms of the vectors they multiply. Specifically,

$$\|Qz\|_2^2 = (zQ)^T(Qz) = z^T(Q^TQ)z = z^Tz = \|z\|^2.$$

Vector and matrix norms that do not change under orthogonal transformations are called *unitarily invariant norms.*

21. The **QR** *decomposition* is a unitary reduction of a matrix to triangular form.

> Let $X$ be an $n \times k$ matrix of rank $k$. Then there is an orthogonal matrix $Q$ such that
> $$Q^TX = \begin{pmatrix} R \\ 0 \end{pmatrix}, \tag{7.5}$$
> where $R$ is an upper triangular matrix with positive diagonal elements.

We will give a constructive proof of the existence of the **QR** decomposition later (§8.9). But first we will show how it can be used to solve least squares problems.

22. Let $Q$ be partitioned in the form

$$Q = (Q_X \ Q_\perp),$$

where $Q_X$ has $k$ columns. Then on multiplying (7.5) by $Q$ we get

$$X = (Q_X \ Q_\perp)\begin{pmatrix} R \\ 0 \end{pmatrix} = Q_X R.$$

From this equation we can derive several facts.

The factorization $X = Q_X R$ is the **QR** factorization of $X$ (§5.21). Consequently, this part of the **QR** decomposition is unique. Moreover, the columns of $Q_X$ form an orthonormal basis for the column space of $X$, and the matrix

$$P_X = Q_X Q_X^T$$

is the projection onto that space (see §6.6).

We also have

$$I = (Q_X \ Q_\perp)(Q_X \ Q_\perp)^T = Q_X Q_X^T + Q_\perp Q_\perp^T = P_X + Q_\perp Q_\perp^T.$$

Hence

$$Q_\perp Q_\perp^T = I - P_X = P_\perp$$

is the projection onto the orthogonal complement of the column space of $X$. Thus the **QR** decomposition gives us an alternate expression for $P_\perp$.

23.   Since the QR decomposition subsumes the QR factorization, it can be used to solve least squares problems via the QR equation (6.9). However, it can also be used directly to rederive the QR equation in a way that gives us something extra.

From the orthogonality of $Q$ we have

$$
\begin{aligned}
\|y - Xb\|_2^2 &= \|Q^{\mathrm{T}}(y - Xb)\|_2^2 \\
&= \left\| \begin{pmatrix} Q_X^{\mathrm{T}} \\ Q_\perp^{\mathrm{T}} \end{pmatrix} y - \begin{pmatrix} R \\ 0 \end{pmatrix} b \right\|_2^2 \\
&= \left\| \begin{pmatrix} z_X - Rb \\ z_\perp \end{pmatrix} \right\|_2^2,
\end{aligned}
\tag{7.6}
$$

where

$$
z_X = Q_X^{\mathrm{T}} y \quad \text{and} \quad z_\perp = Q_\perp^{\mathrm{T}} y.
$$

Now the norm of the last expression in (7.6) is the sum of the squares of the components of $z_X - Rb$ and of $z_\perp$, i.e., the sum of $\|z_X - Rb\|_2^2$ and $\|z_\perp\|_2^2$. Hence

$$
\|y - Xb\|_2^2 = \|z_X - Rb\|_2^2 + \|z_\perp\|_2^2.
\tag{7.7}
$$

The second expression on the right-hand side of (7.7) is independent of $b$. Hence $\|y - Xb\|_2^2$ will be minimized when $z_X - Rb = 0$ — that is, when $b = R^{-1} z_X$ (remember that $R$ is nonsingular). In this case the residual sum of squares is $\|z_\perp\|_2^2$. Moreover, since $Xb = P_X y$ and $y - Xb = P_\perp y$, we have $Xb = Q_X z_X$ and $y - Xb = Q_\perp z_\perp$. Thus the QR decomposition provides a complete solution of the least squares problem with new expressions for the residual and its norm. The facts are summarized in Figure 7.1.

24.   It is instructive to compare the partition of the residual sum of squares

$$
\|y - Xb\|_2^2 = \|z_X - Rb\|_2^2 + \|z_\perp\|_2^2
\tag{7.8}
$$

with the partition

$$
\|y - Xb\|_2^2 = \|P_X y - Xb\|_2^2 + \|P_\perp y\|_2^2,
\tag{7.9}
$$

which was used in §6.11 to derive the QR equation. The second partition decomposes the residual sum of squares into norms of projections onto the column space of $X$ and its orthogonal complement. The first partition can be obtained from the second by transforming everything by $Q^{\mathrm{T}}$, which causes the column space of $X$ to be spanned by the first $k$ unit vectors, $\mathbf{e}_1, \ldots, \mathbf{e}_k$ and the orthogonal complement to be spanned by $\mathbf{e}_{k+1}, \ldots, \mathbf{e}_n$. This repositioning of the column space of $X$ accounts for the computationally simpler form of the first partition.

Let $X \in \mathbf{R}^{n \times k}$ have rank $k$ and let $y \in \mathbf{R}^n$. Let the QR decomposition of $X$ be partitioned in the form

$$\begin{pmatrix} Q_X^{\mathrm{T}} \\ Q_\perp^{\mathrm{T}} \end{pmatrix} X = \begin{pmatrix} R \\ 0 \end{pmatrix},$$

where $Q_X$ has $k$ columns, and set

$$z_X = Q_X^{\mathrm{T}} y \quad \text{and} \quad z_\perp = Q_\perp^{\mathrm{T}} y.$$

Consider the least squares problem:

$$\text{minimize } \|y - Xb\|_2.$$

1.  The vector $b$ satisfies the QR equation

    $$Rb = z_X.$$

2.  The residual vector is given by

    $$y - Xb = Q_\perp z_\perp,$$

    and its norm by
    $$\|y - Xb\|_2 = \|z_\perp\|_2.$$

3.  The approximation $Xb$ is given by

    $$Xb = Q_X z_X.$$

Figure 7.1. *Summary of least squares and the QR decomposition.*

It should not be thought, however, that one approach is better than the other. The decomposition (7.9) is more general than (7.8) in that it applies with equal facility to continuous and discrete problems. The decomposition (7.8) represents a matrix approach that restricts us to discrete problems. However, the loss of generality is compensated by more computationally oriented formulas. You would be well advised to become equally familiar with both approaches.

# Approximation

Householder Transformations
Orthogonal Triangularization
Implementation
Comments on the Algorithm
Solving Least Squares Problems

## Householder transformations

1. In many least squares problems $n$ is very large and much greater than $k$. In such cases, the approach through the QR decomposition seems to have little to recommend it, since the storage of $Q$ requires $n^2$ storage locations. For example, if $n$ is one million, $Q$ requires one trillion floating-point words to store. Only members of Congress, crazed by the scent of pork, can contemplate such figures with equanimity.

2. The cure for this problem is to express $Q$ in a factored form that is easy to store and manipulate. The factors are called Householder transformations. Specifically, a *Householder transformation* is a matrix of the form

$$H = I - uu^{\mathrm{T}}, \qquad \|u\|_2 = \sqrt{2}. \tag{8.1}$$

It is obvious that $H$ is symmetric. Moreover,

$$\begin{aligned} H^{\mathrm{T}}H &= (I - uu^{\mathrm{T}})(I - uu^{\mathrm{T}}) \\ &= I - 2uu^{\mathrm{T}} + (uu^{\mathrm{T}})(uu^{\mathrm{T}}) \\ &= I - 2uu^{\mathrm{T}} + u(u^{\mathrm{T}}u)u^{\mathrm{T}} \\ &= I - 2uu^{\mathrm{T}} + 2uu^{\mathrm{T}} \\ &= I, \end{aligned}$$

so that $H$ is orthogonal—a good start, since we want to factor $Q$ into a sequence of Householder transformations.

3. The storage of Householder transformations presents no problems, since all we need is to store the vector $u$. However, this ploy will work only if we can manipulate Householder transformations in the form $I - uu^{\mathrm{T}}$. We will now show how to premultiply a matrix $X$ by a Householder transformation $H$.
From (8.1)

$$HX = (I - uu^{\mathrm{T}})X = X - u(u^{\mathrm{T}}X) = X - uv^{\mathrm{T}},$$

where $v^{\mathrm{T}} = u^{\mathrm{T}}X$. Thus we have the following algorithm.

$$1. \quad v^{\mathrm{T}} = u^{\mathrm{T}} X$$
$$2. \quad X = X - uv^{\mathrm{T}} \tag{8.2}$$

It is easy to verify that if $X \in \mathbf{R}^{n \times k}$, then (8.2) requires about $2kn$ multiplications and additions. This should be compared with a count of $kn^2$ for the explicit multiplication by a matrix of order $n$. What a difference the position of the constant two makes!

4. Although Householder transformations are convenient to store and manipulate, we have yet to show they are good for anything. The following result remedies this oversight by showing how Householder transformations can be used to introduce zeros into a vector.

Suppose $\|x\|_2 = 1$ and let

$$u = \frac{x \pm \mathbf{e}_1}{\sqrt{1 \pm \xi_1}}.$$

If $u$ is well defined, then

$$Hx \equiv (I - uu^{\mathrm{T}})x = \mp \mathbf{e}_1.$$

We first show that $H$ is a Householder transformation by showing that $\|u\|_2 = \sqrt{2}$. Specifically,

$$
\begin{aligned}
\|u\|_2^2 &= \frac{(x \pm \mathbf{e}_1)^{\mathrm{T}}(x \pm \mathbf{e}_1)}{1 \pm \xi_1} \\
&= \frac{x^{\mathrm{T}}x \pm 2\mathbf{e}_1^{\mathrm{T}}x + \mathbf{e}_1^{\mathrm{T}}\mathbf{e}_1}{1 \pm \xi_1} \\
&= \frac{1 \pm 2\xi_1 + 1}{1 \pm \xi_1} \\
&= 2.
\end{aligned}
$$

Now $Hx = x - (u^{\mathrm{T}}x)u$. But

$$
\begin{aligned}
u^{\mathrm{T}}x &= \frac{(x \pm \mathbf{e}_1)^{\mathrm{T}}x}{\sqrt{1 \pm \xi_1}} \\
&= \frac{x^{\mathrm{T}}x \pm \mathbf{e}_1^{\mathrm{T}}x}{\sqrt{1 \pm \xi_1}} \\
&= \frac{1 \pm \xi_1}{\sqrt{1 \pm \xi_1}} \\
&= \sqrt{1 \pm \xi_1}.
\end{aligned}
$$

Hence

$$Hx = x - (u^\mathrm{T}x)u$$
$$= x - \sqrt{1 \pm \xi_1}\frac{x \pm \mathbf{e}_1}{\sqrt{1 \pm \xi_1}}$$
$$= x - (x \pm \mathbf{e}_1)$$
$$= \mp\mathbf{e}_1.$$

5. The proof shows that there are actually two Householder transformations that transform $x$ into $\mp\mathbf{e}_1$. They correspond to the two choices of the sign $\sigma = \pm 1$ in the construction of $u$. Now the only case in which this construction can fail is when $|\xi_1| = 1$ and we choose the sign so that $1 + \sigma\xi_1 = 0$. We can avoid this case by always choosing $\sigma$ to have the same sign as $\xi_1$.

This choice of $\sigma$ doesn't just prevent breakdown; it is essential for numerical stability. If $\xi_1$ is near one and we choose the wrong sign, then cancellation will occur in the computation $1 + \sigma\xi_1$.[9]

6. A variant of this construction may be used to reduce any nonzero $x$ vector to a multiple of $\mathbf{e}_1$. Simply compute $u$ from $x/\|x\|_2$. In this case

$$Hx = \mp\|x\|_2\mathbf{e}_1.$$

The following algorithm implements this procedure. In addition to returning $u$ it returns $\nu = \mp\|x\|_2$.

```
1.   housegen(x, u, ν)
2.      ν = ‖x‖₂
3.      if (ν = 0) u = √2e₁; return; fi
4.      u = x/ν
5.      if (u₁ ≥ 0)
6.         σ = 1
7.      else
8.         σ = -1
9.      end if
10.     u₁ = u₁ + σ
11.     u = u/√|u₁|
12.     ν = -σν
13.  end housegen
```

## Orthogonal triangularization

7. The equation $Hx = \mp\|x\|_2\mathbf{e}_1$ is essentially a QR decomposition of $X$ with $H$ playing the role of $Q$ and $\mp\|x\|_2$ playing the role of $R$. (We add

---

[9]There is, however, an alternate, stable formula. See Å. Björck, *Numerical Methods for Least Squares,* SIAM, 1996, p. 53.

$$
\begin{array}{cccc}
x & x & x & x \\
\hat{x} & x & x & x \\
\hat{x} & x & x & x \\
\hat{x} & x & x & x \\
\hat{x} & x & x & x \\
\hat{x} & x & x & x
\end{array}
\quad \overset{H_1}{\Longrightarrow} \quad
\begin{array}{cccc}
r & r & r & r \\
0 & x & x & x \\
0 & \hat{x} & x & x \\
0 & \hat{x} & x & x \\
0 & \hat{x} & x & x \\
0 & \hat{x} & x & x
\end{array}
\quad \overset{H_2}{\Longrightarrow} \quad
\begin{array}{cccc}
r & r & r & r \\
0 & r & r & r \\
0 & 0 & x & x \\
0 & 0 & \hat{x} & x \\
0 & 0 & \hat{x} & x \\
0 & 0 & \hat{x} & x
\end{array}
\quad \overset{H_3}{\Longrightarrow}
$$

$$
\overset{H_3}{\Longrightarrow} \quad
\begin{array}{cccc}
r & r & r & r \\
0 & r & r & r \\
0 & 0 & r & r \\
0 & 0 & 0 & x \\
0 & 0 & 0 & \hat{x} \\
0 & 0 & 0 & \hat{x}
\end{array}
\quad \overset{H_4}{\Longrightarrow} \quad
\begin{array}{cccc}
r & r & r & r \\
0 & r & r & r \\
0 & 0 & r & r \\
0 & 0 & 0 & r \\
0 & 0 & 0 & 0 \\
0 & 0 & 0 & 0
\end{array}
$$

Figure 8.1. *Orthogonal triangularization.*

the qualification "essentially" because the diagonal of $R = (\mp\|x\|_2)$ may be negative.) We are now going to show how to parlay this minidecomposition into the QR decomposition of a general $n \times k$ matrix $X$. We will look at the process in two ways — schematically and recursively — and then write a program.

8. The schematic view is illustrated in Figure 8.1 for $n = 6$ and $k = 4$. The arrays in the figure are called Wilkinson diagrams (after the founding father of modern matrix computations). A letter in a diagram represents an element which may or may not be zero (but is usually nonzero). A zero in the diagram represents an element that is definitely zero.

The scheme describes a reduction to triangular form by a sequence of Householder transformations. At the first step, the first column is singled out and a Householder transformation $H_1$ is generated that reduces the column to a multiple of $\mathbf{e}_1$ — i.e., it replaces the elements marked with hats by zeros. The transformation is then multiplied into the *entire* matrix to get the second array in the figure.

At the second step, we focus on the trailing $5 \times 3$ submatrix, enclosed by a box in the figure. We use the first column of the submatrix to generate a Householder transformation $H_2$ that annihilates the elements with hats and multiply $H_2$ into the submatrix to get the next array in the sequence.

The process continues until we run out of columns. The result is an upper triangular matrix $R$ (hence the $r$'s in the Wilkinson diagrams). More precisely, we have

$$
\begin{pmatrix} I_3 & 0 \\ 0 & H_4 \end{pmatrix}
\begin{pmatrix} I_2 & 0 \\ 0 & H_3 \end{pmatrix}
\begin{pmatrix} I_1 & 0 \\ 0 & H_2 \end{pmatrix}
H_1 X = \begin{pmatrix} R \\ 0 \end{pmatrix}.
$$

Since the product of orthogonal transformations is orthogonal, the $Q^\mathrm{T}$ of our QR decomposition is given by the product

$$Q^\mathrm{T} = \begin{pmatrix} I_3 & 0 \\ 0 & H_4 \end{pmatrix} \begin{pmatrix} I_2 & 0 \\ 0 & H_3 \end{pmatrix} \begin{pmatrix} I_1 & 0 \\ 0 & H_2 \end{pmatrix} H_1.$$

9.     The recursive approach, to which we now turn, is a standard way of deriving matrix algorithms. It is also a formal proof of the existence of the QR decomposition.

Partition $X$ in the form

$$X = (x_{11} \ X_{12}),$$

where we use northwest indexing. (See §6.1. Note that $x_{11}$ is a vector.) Determine a Householder transformation $H_1$ so that

$$H_1 x_{11} = \begin{pmatrix} \rho_{11} \\ 0 \end{pmatrix}.$$

Then $H_1 X$ has the form

$$H_1 X = \begin{pmatrix} \rho_{11} & r_{12}^\mathrm{T} \\ 0 & \hat{X}_{22} \end{pmatrix}.$$

Now by an obvious induction, the matrix $\hat{X}_{22}$ also has a QR decomposition:

$$\hat{Q}_{22}^\mathrm{T} \hat{X}_{22} = \begin{pmatrix} R_{22} \\ 0 \end{pmatrix}.$$

If we set

$$Q^\mathrm{T} = \begin{pmatrix} 1 & 0 \\ 0 & \hat{Q}_{22}^\mathrm{T} \end{pmatrix} H_1,$$

then

$$Q^\mathrm{T} X = \begin{pmatrix} 1 & 0 \\ 0 & \hat{Q}_{22} \end{pmatrix} H_1 X = \begin{pmatrix} 1 & 0 \\ 0 & \hat{Q}_{22} \end{pmatrix} \begin{pmatrix} \rho_{11} & r_{12}^\mathrm{T} \\ 0 & \hat{X}_{22} \end{pmatrix} = \begin{pmatrix} \rho_{11} & r_{12}^\mathrm{T} \\ 0 & R_{22} \\ 0 & 0 \end{pmatrix}.$$

Hence if we write

$$R = \begin{pmatrix} \rho_{11} & r_{12}^\mathrm{T} \\ 0 & R_{22} \end{pmatrix},$$

then

$$Q^\mathrm{T} X = \begin{pmatrix} R \\ 0 \end{pmatrix}$$

is the required decomposition. (Well, almost. The diagonals of $R$ can be negative. It is a good exercise to figure out what can be done about this and to decide if it is worth doing.)

## Implementation

10.  Even when they are based on a recursive derivation, matrix algorithms typically do not use recursion. The reason is not that the overhead for recursion is unduly expensive — it is not. Rather, many matrix algorithms, when they fail, do so at points deep inside the algorithm. It is easier to jump out of a nest of loops than to swim up from the depths of a recursion. Thus, however we derive a matrix algorithm, we are faced with the problem of providing a direct implementation.

11.  One way of doing this is to imagine what the matrix looks like at the $j$th stage of the algorithm and write down the partitioned form in northeast indexing. This can be done easily from a Wilkinson diagram and with a little practice from the recursive derivation. For orthogonal triangularization, at the $j$th stage the matrix has the form

$$\begin{pmatrix} R_{11} & r_{1j} & R_{1,j+1} \\ 0 & x_{jj} & X_{j,j+1} \end{pmatrix}, \tag{8.3}$$

where the first row represents the part of the matrix that will not change.

12.  Before we can implement our algorithm, we must decide where we will store the objects we are computing. Initially we will be profligate in our use of storage and will economize later on. Specifically, we will store the matrix $X$, the Householder vectors $u$, and the matrix $R$ in separate arrays called $X$, $U$, and $R$.

13.  We will now write down the steps of the derivation and in parallel the corresponding code. Because we have used northwest indexing in the partition (8.3) we can easily write down the correspondence between the matrix $X$ and the array that contains it:

$$x_{jj}, \ X_{j,j+1} \ \longleftrightarrow \ X[j{:}n, j], \ X[j{:}n, j{+}1{:}k]$$

The first step at the $j$th stage is to

$$\begin{array}{l}\text{generate the} \\ \text{Householder vector } u_j\end{array} \ \longleftrightarrow \ housegen(X[j{:}n, j], U[j{:}n, j], R[j, j])$$

Note that we have stored the Householder vector in the same position in the array $U$ as the vector $x_{jj}$ occupies in the array $X$. The norm of $x_{jj}$, with an appropriate sign, is the element $\rho_{jj}$ of $R$ and is moved there forthwith.

The next step is to compute $H_j X_{j,j+1} = (I - u_j u_j^{\mathrm{T}}) X_{j,j+1}$. This is done in two stages as shown in (8.2):

$$v^{\mathrm{T}} = u_j^{\mathrm{T}} X_{j,j+1} \ \longleftrightarrow \ v^{\mathrm{T}} = U[j{:}n, j]^{\mathrm{T}} * X[j{:}n, j{+}1, k]$$

and

$$X_j - u_j v^{\mathrm{T}} \longleftrightarrow X[j{:}n, j{+}1{:}k] = X[j{:}n, j{+}1{:}k] - U[j{:}n, j]*v^{\mathrm{T}}$$

Finally, we

$$\text{update } R \longleftrightarrow R[j, j{+}1{:}k] = X[j, j{+}1{:}k]$$

14. The following algorithm combines the above steps.

```
1.   QRD(X, U, R)
2.      for j = 1 to k
3.         housegen(X[j:n, j], U[j:n, j], R[j, j])
4.         v[j+1:k]ᵀ = U[j:n, j]ᵀ*X[j:n, j+1, k]
5.         X[j:n, j+1:k] = X[j:n, j+1:k] − U[j:n, j]*v[j+1:k]ᵀ
6.         R[j, j+1:k] = X[j, j+1:k]
7.      end for j
8.   end QRD
```

## Comments on the algorithm

15. The program $QRD$ has been coded at a rather high level. People who encounter this style of programming for the first time often feel that it has to be inefficient. Wouldn't it be better (they ask) to write out the loops?

In fact, this high-level coding is potentially more efficient. The matrix-vector operations in the individual statements can be assigned to a library of general-purpose subroutines, which then can be optimized for the machine in question. Although the initial effort is great, it is compensated for by faster running times. Subroutines of this kind are known in the trade as BLAS — basic linear algebra subprograms.

16. A second advantage of the high-level approach is that we can introduce special conventions to handle conditions at the beginning and end of a loop. For example, when $j = k$, statements 4 and 5 are not needed, since there is no submatrix to the right of the last column. If we adopt the convention that inconsistent index ranges like $k{+}1{:}k$ suppress execution, then these statements will be ignored when $j = k$. These conventions also take care of the case $n \leq k$.

17. We have been very free in our use of storage. In many applications, however, the matrix $X$ is not needed after its QR decomposition has been computed. In this case it makes sense to overwrite $X$ with the Householder vectors $u_i$. In $QRD$ this can be done by everywhere replacing the variable $U$ with $X$.

Industrial strength codes, such as those found in LINPACK and LAPACK, actually store both $U$ and $R$ in the array $X$. There is a conflict here, since both require the main diagonal. It is resolved by storing the first elements of the vectors $u_i$ in a separate array.

18. Although the number of arithmetic operations required by an algorithm is
not a reliable measure of its running time, such counts are useful in comparing
algorithms. The bulk of the arithmetic work is done in statements 4 and 5,
which are readily seen to perform about $2(n-j)(k-j)$ operations. Since $j$
ranges from 1 to $k$, the total count is

$$2\sum_{j=1}^{k}(n-j)(k-j) \cong 2\int_{0}^{k}(n-j)(k-j)\,dj = nk^2 - \frac{1}{3}k^3. \qquad (8.4)$$

## Solving least squares problems

19. The quantities computed by $QRD$ can be used to solve least squares
problems. The process involves the successive applications of the Householder
transformations generated in the course of the reduction. In terms of the
quantities computed by the code, the product of the $j$th transformation with
a vector $y$ can be implemented as follows.

1. $temp = U[j{:}n, j]^{\mathrm{T}} y[j{:}n]$
2. $y[j{:}n] = y[j{:}n] - temp*U[j{:}n, j]$ $\qquad (8.5)$

This operation, which we will write symbolically as $H_j y$, requires about $2(n-j)$
additions and multiplications.

20. The natural approach to solving least squares problems via the **QR** decomposition is to form and solve the **QR** equation. From Figure 7.1 we see
that the first thing we must do is compute

$$\begin{pmatrix} z_X \\ z_\perp \end{pmatrix} = Q^{\mathrm{T}} y.$$

Since $Q^{\mathrm{T}} = H_k \cdots H_1$, we can perform this computation by $k$ applications of
(8.5).

21. Once we have $z_X$, we can solve the triangular system

$$Rb = z_X$$

for the least squares solution $b$.

22. The residual sum of squares can be calculated in the form $\|z_\perp\|_2^2$. If we
have overwritten $X$ with the Householder vectors, we cannot calculate the
residual in the form $y - Xb$. However, we have the alternative form

$$y - Xb = Q_\perp z_\perp = (Q_Z \; Q_\perp)\begin{pmatrix} 0 \\ z_\perp \end{pmatrix} = H_1 \cdots H_k \begin{pmatrix} 0 \\ z_\perp \end{pmatrix}. \qquad (8.6)$$

Thus we can calculate the residual by repeated application of (8.5).

Similarly, we can calculate the approximation $Xb$ in the form

$$Xb = H_1 \cdots H_k \begin{pmatrix} z_X \\ 0 \end{pmatrix}. \tag{8.7}$$

23.   In some applications we need to compute the projections $P_X y$ and $P_\perp y$ of a vector $y$ without reference to least squares approximation. Now $P_X y$ is just the least squares approximation, which can be calculated as in (8.7). Similarly, $P_\perp y$ is the least squares residual, which can be calculated as in (8.6).

# Approximation

Operation Counts
The Frobenius and Spectral Norms
Stability of Orthogonal Triangularization
Error Analysis of the Normal Equations
Perturbation of Inverses and Linear Systems
Perturbation of Pseudoinverses and Least Squares Solutions
Summary

## Operation counts

1. We now have two methods for solving least squares problems — the method of normal equations and the method of orthogonal triangularization. But which should we use? Unfortunately, there is no simple answer. In this lecture, we will analyze three aspects of these algorithms: speed, stability, and accuracy.

2. We have already derived an operation count (8.4) for the method of orthogonal triangularization. If $n$ is reasonably greater than $k$ then the computation of the QR decomposition by Householder transformations requires about

$$nk^2 \text{ additions and multiplications.} \tag{9.1}$$

3. Under the same assumptions, the solution by the normal equations requires about

$$\tfrac{1}{2}nk^2 \text{ additions and multiplications.} \tag{9.2}$$

To see this let $x_j$ denote the $j$th column of $X$. Then the $(i,j)$-element of $A = X^{\mathrm{T}}X$ is

$$a_{ij} = x_i^{\mathrm{T}}x_j.$$

This inner product requires about $n$ additions and multiplications for its formation. By symmetry we only have to form the upper or lower half of the matrix $A$. Thus we must calculate $\tfrac{1}{2}k^2$ elements at a cost of $n$ additions and multiplications per element.

  Once the normal equations have been formed, their solution by the Cholesky algorithm requires $\tfrac{1}{6}k^3$ additions and multiplications. Since in least squares applications $n$ is greater than $k$ — often much greater — the bulk of the work is in the formation of the normal equations, not in their solution.

4. Comparing (9.1) and (9.2), we see that the method of normal equations is superior. Moreover, in forming the cross-product matrix $X^{\mathrm{T}}X$, we can often take advantage of special structure in the matrix $X$, which is harder to do in calculating the QR decomposition. Thus, if our only concern is speed, the normal equations win hands down.

## The Frobenius and spectral norms

5.   We are going to bound errors associated with the algorithms under investigation. Since some of these errors are in matrices like $X$ or $A$, we will need to extend the notion of norm to matrices. Rather than give a general treatment, we will introduce two norms that are widely used in numerical analysis — particularly in least squares — and list their properties.

The norms are the Frobenius norm and the spectral norm. The Frobenius norm $\|\cdot\|_{\mathrm{F}}$ is the natural generalization of the Euclidean vector norm:

$$\|X\|_{\mathrm{F}}^2 = \sum_{i,j} x_{ij}^2.$$

The spectral norm (or matrix 2-norm) is defined by

$$\|X\|_2 = \max_{\|b\|_2=1} \|Xb\|_2.$$

When $X$ is a vector — i.e., when it has only one column — both norms reduce to the usual vector 2-norm.

6.   The Frobenius norm is widely used because it is easy to compute. The spectral norm is expensive to compute, but since it is always smaller than the Frobenius norm, it generally gives sharper bounds.

7.   Both norms have the following properties, whose verification is left as an exercise.

1.   They are defined for matrices of all dimensions.
2.   They are norms on $\mathbf{R}^{n \times k}$; that is,

   1.   $X \neq 0 \implies \|X\| > 0$,
   2.   $\|\alpha X\| = |\alpha| \|X\|$,
   3.   $\|X + Y\| \leq \|X\| + \|Y\|$.

3.   They are *consistent* in the sense that

$$\|XY\| \leq \|X\| \|Y\|,$$

   whenever the product $XY$ is defined.
4.   They are symmetric; that is,

$$\|X^{\mathrm{T}}\| = \|X\|.$$

5.   The norm of the identity matrix is not less than one.

In what follows we will use the symbol $\| \cdot \|$ to stand for any norm with the above properties.

8. If $X$ is a matrix and $\tilde{X}$ is an approximation to $X$ we can define the *error* in $\tilde{X}$ by

$$\| \tilde{X} - X \|$$

and the *relative error* in $\tilde{X}$ by

$$\frac{\| \tilde{X} - X \|}{\| X \|}.$$

These definitions are formed in analogy with the usual definition of error and relative error for real numbers. However, because they summarize the status of an array of numbers in a single quantity they do not tell the whole story. In particular, they do not provide much information about the smaller components of $\tilde{X}$. For example, let

$$x = \begin{pmatrix} 1 \\ 10^{-5} \end{pmatrix} \quad \text{and} \quad \tilde{x} = \begin{pmatrix} 1.00001 \\ 2 \cdot 10^{-5} \end{pmatrix}.$$

Then the vector $\tilde{x}$ has a relative error of about $10^{-5}$ in the 2-norm. The first component of $\tilde{x}$ has the same relative error. But the second component, which is small, has a relative error of one! You should keep this example in mind in interpreting the bounds to follow.

## Stability of orthogonal triangularization

9. We now turn to the effects of rounding error on the method of orthogonal triangularization. We will assume that the computations are carried out in floating-point arithmetic with rounding unit $\epsilon_M$. If, for example, we are performing double-precision computations in IEEE standard arithmetic we will have $\epsilon_M \cong 10^{-16}$, sixteen being the roughly the number of significant digits carried in a double-precision word.

10. Since the rounding-error analysis of orthogonal triangularization is quite involved, we will only state the final results.

Let the solution of the least squares problem of minimizing $\|y - Xb\|_2$ be computed by orthogonal triangularization in floating-point arithmetic with rounding unit $\epsilon_M$, and let $\tilde{b}$ denote the result. Then there is a matrix $E$ and vector $f$ satisfying

$$\frac{\|E\|}{\|X\|}, \frac{\|f\|}{\|y\|} \le c_{nk}\epsilon_M \qquad (9.3)$$

such that

$$\|(y + f) - (X + E)\tilde{b}\|_2 \text{ is minimized.} \qquad (9.4)$$

Here $c_{nk}$ is a slowly growing function of $n$ and $k$. The same result holds for the computed approximation $Xb$ and residual $y - Xb$, although the matrix $E$ will be different in each case.

11. The result just cited projects the rounding errors made in the computation back on the original matrix $X$ and the vector $y$. For this reason, the solution of least squares problems by orthogonal triangularization is said to be *backward stable* (or simply stable when the context is clear). Backward stability has remarkable explanatory powers.

Suppose, for example, that someone complains to you about the quality of a least squares solution computed by orthogonal triangularization in double precision. The first thing to do is to ask how accurate the matrix $X$ is. Suppose the response is that the matrix is accurate to six decimal digits. Then you can immediately respond that the computation was equivalent to introducing errors in, say, the fourteenth or fifteenth digits of $X$, errors which are completely dominated by the errors that are already present. The original errors, not the algorithm, are responsible for the unsatisfactory solution.

12. While this response may be satisfying to you, it is not likely to be well received by your customer, who will immediately ask what effect the errors — rounding or otherwise — have on the solution. This is a problem in perturbation theory. We will treat it a little later, after we have looked at the rounding-error analysis of the normal equations.

### Error analysis of the normal equations

13. To solve least squares problems via the normal equations we form the cross-product matrix

$$A = X^{\mathrm{T}}X$$

and then solve the system

$$Ab = X^{\mathrm{T}}y.$$

The effects of rounding error on this procedure are as follows.

Let the solution of the least squares problem of minimizing $\|y-Xb\|_2$ be computed by the normal equations in floating-point arithmetic with rounding unit $\epsilon_M$, and let $\tilde{b}$ denote the result. Then

$$(A + G)\tilde{b} = X^T y, \qquad (9.5)$$

where

$$\frac{\|G\|}{\|A\|} \le d_{nk}\left(1 + \frac{\|y\|}{\|X\|\|\tilde{b}\|}\right)\epsilon_M. \qquad (9.6)$$

Here $d_{nk}$ is a slowly growing function of $n$ and $k$.

14. In most applications, $X\tilde{b}$ will be a ballpark approximation to $y$, so that $\|y\|/\|X\|\|\tilde{b}\|$ will be less than one. Thus the bound (9.6) is a slowly growing multiple of $\epsilon_M$.

15. The inequality (9.6) is a backward error bound in the sense that the effects of the rounding error are projected back on the normal equations. But it is less satisfactory than the bound for orthogonal triangularization because it does not project the error back on the original data $X$. It is sometimes possible to throw the error in $A$ all the way back to $X$, but this latter error will always be greater than the error $E$ for orthogonal triangularization — sometimes much greater.

16. Another way of looking at the same thing is that the rounded normal equations do not preserve as much information as the rounded $X$ and $y$. For example, suppose

$$X = \begin{pmatrix} 1 \\ 3.1415\cdot 10^{-6} \end{pmatrix}.$$

If $X^T X$ is rounded to ten decimal digits, the result is a matrix consisting of the number one. All information about the number $3.1415\cdot 10^{-6}$ has been lost in the passage to the cross-product matrix. If those digits are important, we are in trouble.

17. Thus, on grounds of stability the orthogonal triangularization beats the normal equations. The score is now tied. Let's look at accuracy.

## Perturbation of inverses and linear systems

18. To get more insight into the difference between orthogonal triangularization and the normal equations, we will investigate the effects of the perturbation $E$ in (9.4) and $G$ in (9.5) on the solution of the least squares solutions. We will begin with the normal equations because the analysis is simpler.

19. The solution of the system $(A + G)\tilde{b} = c$ can be written in the form

$$\tilde{b} = (A + G)^{-1}c.$$

If we set $(A + G)^{-1} = A^{-1} + H$, then

$$\tilde{b} - b = Hc.$$

Thus the problem of determining the error in $\tilde{b}$ can be reduced to the problem of determining $H$, the error in $(A + G)^{-1}$ as an approximation to $A^{-1}$.

20.   Instead of computing a rigorous bound on $H$, we will use a technique called *first-order perturbation theory* to compute an approximation to $H$. The underlying idea is simple. Since the inverse is differentiable, the quantity $H$ must be of the same order as $G$. Consequently, products like $GH$ go to zero faster that either $H$ or $G$ and hence can be ignored.

21.   We begin with an equation that says $A^{-1} + H$ is the inverse of $(A + G)$: namely,

$$(A + G)(A^{-1} + H) = I.$$

Expanding this equation and simplifying, we get

$$AH + GA^{-1} + GH = 0.$$

If we now ignore the quantity $GH$, we obtain the approximate equation $AH + GA^{-1} \cong 0$, which can be solved for $H$ to give

$$H \cong -A^{-1}GA.$$

The error in this approximation will be proportional to the size of the terms we have ignored, which is $GH = O(\|G\|^2)$. The following summarizes this result.

---

Let $A$ be nonsingular. Then for all sufficiently small $G$,

$$(A + G)^{-1} = A^{-1} - A^{-1}GA^{-1} + O(\|G\|^2). \qquad (9.7)$$

---

22.   Returning now to the system $(A + G)\tilde{b} = c$, we have from (9.7) that

$$\tilde{b} - b \cong -A^{-1}GA^{-1}c = -A^{-1}Gb.$$

Hence

$$\frac{\|\tilde{b} - b\|}{\|b\|} \lesssim \|A^{-1}G\|.$$

By weakening this bound, we can write it in a form that is easy to interpret. Specifically, let $\kappa(A) = \|A\|\|A^{-1}\|$. Then

$$\frac{\|\tilde{b} - b\|}{\|b\|} \lesssim \|A^{-1}\|\|G\| = \kappa(A)\frac{\|G\|}{\|A\|}.$$

More precisely:

> Let $A$ be nonsingular and let $Ab = c$. Then for all sufficiently small $G$, the solution of the equation $(A + G)\tilde{b} = c$ satisfies
>
> $$\frac{\|\tilde{b} - b\|}{\|b\|} \leq \kappa(A)\frac{\|G\|}{\|A\|} + O(\|G\|^2), \qquad (9.8)$$
>
> where
> $$\kappa(A) = \|A\|\|A^{-1}\|.$$

23.   The number $\kappa(A)$ is called *the condition number of $A$ with respect to inversion.* It is greater than one, since

$$1 \leq \|I\| = \|AA^{-1}\| \leq \|A\|\|A^{-1}\| = \kappa(A).$$

Now the right-hand side of (9.8) is the relative error in $\tilde{b}$ as an approximation to $b$. The quantity $\|G\|/\|A\|$ is the relative error in $A+G$ as an approximation to $A$. Hence the condition number shows how a relative error in the matrix $A$ is magnified in the solution of the system $Ab = c$.

24.   Returning to the least squares problem and the normal equations, we see from (9.5) that the computed solution $\tilde{b}$ satisfies

$$\frac{\|\tilde{b} - b\|}{\|b\|} \lesssim d_{nk}\kappa(A)\left(1 + \frac{\|y\|}{\|X\|\|\tilde{b}\|}\right)\epsilon_{\mathrm{M}}.$$

## Perturbation of pseudoinverses and least squares solutions

25.   We now wish to investigate the effects of the perturbation $E$ on the vector $\tilde{b}$ that minimizes $\|y - (X + E)b\|_2$. Just as we approached perturbation theory for linear systems by way of the inverse matrix — i.e., the matrix that produces the solution — we will approach the perturbation of least squares solution by way of a matrix called the pseudoinverse.

26.   Let $X$ have linearly independent columns. The *pseudoinverse* of $X$ is the matrix

$$X^{\dagger} = (X^{\mathrm{T}}X)^{-1}X^{\mathrm{T}}.$$

The pseudoinverse has the following easily verified properties.

$$\begin{aligned}
&1.\quad b = X^{\dagger}y \text{ minimizes } \|y - Xb\|_2, \\
&2.\quad X^{\dagger}X = I, \\
&3.\quad XX^{\dagger} = (X^{\dagger})^{\mathrm{T}}X^{\mathrm{T}} = P_X, \\
&4.\quad X^{\dagger}(X^{\dagger})^{\mathrm{T}} = (X^{\mathrm{T}}X)^{-1}.
\end{aligned} \qquad (9.9)$$

**27.** We will now develop a perturbation expansion for $(X + E)^\dagger$. We begin by expanding $[(X + E)^{\mathrm{T}}(X + E)]^{-1}$. If for brevity we set $A = X^{\mathrm{T}}X$, we have

$$
\begin{aligned}
[(X + E)^{\mathrm{T}}(X + E)]^{-1} &= (A + X^{\mathrm{T}}E + E^{\mathrm{T}}X + E^{\mathrm{T}}E)^{-1} \\
&\cong (A + X^{\mathrm{T}}E + E^{\mathrm{T}}X)^{-1} \qquad \text{ignoring } E^{\mathrm{T}}E \\
&\cong A^{-1} - A^{-1}(E^{\mathrm{T}}X + X^{\mathrm{T}}E)A^{-1} \quad \text{by (9.7)} \\
&= A^{-1} - A^{-1}E(X^\dagger)^{\mathrm{T}} - X^\dagger E A^{-1}.
\end{aligned}
$$

It then follows that

$$
\begin{aligned}
X^\dagger &\cong (A^{-1} - A^{-1}E(X^\dagger)^{\mathrm{T}} - X^\dagger E A^{-1})(X^{\mathrm{T}} + E^{\mathrm{T}}) \\
&\cong A^{-1}X^{\mathrm{T}} + A^{-1}E - A^{-1}E(X^\dagger)^{\mathrm{T}}X^{\mathrm{T}} - X^\dagger E A^{-1}X^{\mathrm{T}} \\
&= X^\dagger A^{-1}E - A^{-1}P_X - X^\dagger E X^\dagger \qquad \text{by (9.9.3)} \\
&= X^\dagger - X^\dagger E X^\dagger + A^{-1}E P_\perp \qquad P_\perp = I - P_X.
\end{aligned}
$$

Thus we have the following result.

> Let $X$ have linearly independent columns. Then for all sufficiently small $E$,
> $$(X + E)^\dagger = X^\dagger - X^\dagger E X^\dagger + A^{-1}E P_\perp + O(\|E\|^2).$$

**28.** It is instructive to ask what happens to this expansion when $X$ is square and nonsingular. In this case $X^\dagger = X^{-1}$. Moreover, since the columns of $X$ span the entire space, we have $P_\perp = 0$. Hence we get $(X + E)^{-1} \cong X^{-1} - X^{-1}EX$, which is exactly (9.7) in a different notation.

**29.** We are now in a position to describe the perturbation of the least squares solution. For simplicity we will ignore the backward error $f$ in (9.3). In what follows we set

$$r = y - Xb = P_\perp y.$$

We have

$$
\begin{aligned}
\tilde{b} &= (X + E)^\dagger y \\
&\cong (X^\dagger - X^\dagger E X^\dagger + A^{-1}E^{\mathrm{T}}P_\perp)y \\
&= b - X^\dagger E b + A^{-1}E^{\mathrm{T}}r.
\end{aligned}
$$

On taking norms and dividing by $\|b\|$, we get

$$\frac{\|\tilde{b} - b\|}{\|b\|} \lesssim \|X^\dagger\|\|E\| + \frac{\|A^{-1}\|\|E\|\|r\|}{\|b\|}. \tag{9.10}$$

But from (9.9.4),

$$\|A^{-1}\| = \|X^\dagger(X^\dagger)^{\mathrm{T}}\| \le \|X^\dagger\|^2. \tag{9.11}$$

Combining (9.10) and (9.11), we have the following result.

---

Let $X$ have linear independent columns. Let $b = X^\dagger y$ and $\tilde{b} = (X + E)^\dagger y$. Then for all sufficiently small $E$,

$$\frac{\|\tilde{b} - b\|}{\|b\|} \leq \left(\kappa(X) + \kappa^2(X)\frac{\|r\|}{\|X\|\|b\|}\right)\frac{\|E\|}{\|X\|} + O(\|E\|^2), \qquad (9.12)$$

where

$$\kappa(X) = \|X\|\|X^\dagger\|.$$

---

30. As with the perturbation of solutions of linear systems, the bound (9.12) shows how relative perturbations in $X$ are magnified in the solution $b$. But the magnification constant or condition number

$$\kappa(X) + \kappa^2(X)\frac{\|r\|}{\|X\|\|b\|}$$

is more complicated. The first term is analogous to the magnification constant for linear systems. But the second term, which depends on $\kappa^2(X)$, is new. Since $\kappa(X) \geq 1$, the second term is potentially much greater than the first. However, it can come into play only if the residual $r$ is large enough.

   To see the effect of the second term (often called the $\kappa^2$ *effect*) consider the matrix

$$X = \begin{pmatrix} 3.0088 & 6.0176 \\ -8.4449 & -16.8904 \\ -1.6335 & -3.2662 \\ -2.1549 & -4.3095 \end{pmatrix}.$$

The value of $\kappa(X)$ is $4.7 \cdot 10^4$. The two vectors[10]

$$y_1 = \begin{pmatrix} 9.0264 \\ -25.3353 \\ -4.8997 \\ -6.4644 \end{pmatrix} \quad \text{and} \quad y_2 = \begin{pmatrix} 33.1046 \\ -17.4267 \\ 2.0116 \\ -9.0774 \end{pmatrix}$$

both have the solution $b = (1\ 1)^\mathrm{T}$. But the first has a residual of zero, while the second has a residual of norm about 26. Figure 9.1 shows the error bounds and the actual errors for a sequence of perturbations $E$. We see two things. First, the bounds are overestimates but not great overestimates. Second, the square of $\kappa(X)$ has a strong influence on the solution $b_2$ corresponding to $y_2$ but no influence on $b_1$, for which the residual $r$ is zero.

---

[10]The components of $x$ and $y_1$ are accurate to about 16 digits; e.g., $x_{11} = 3.00880000000000$ and $y_1^{(1)} = -25.33530000000000$. The components of $y_2$ have nonzero digits beyond those displayed; e.g., $y_1^{(2)} = 33.10464936004095$. It is a good exercise to figure out what would happen if you used the displayed values of $y_2$ instead of their true values.

| $b$ | $\|E\|$ | bound | error |
|---|---|---|---|
|  | 3.2e−05 |  |  |
| 1 |  | 7.2e−02 | 4.8e−02 |
| 2 |  | 3.0e+03 | 1.8e+03 |
|  | 5.0e−06 |  |  |
| 1 |  | 1.1e−02 | 2.3e−03 |
| 2 |  | 4.8e+02 | 3.2e+02 |
|  | 5.3e−07 |  |  |
| 1 |  | 1.2e−03 | 6.1e−04 |
| 2 |  | 5.0e+01 | 2.0e+01 |
|  | 4.2e−08 |  |  |
| 1 |  | 9.6e−05 | 5.5e−05 |
| 2 |  | 4.0e+00 | 5.3e−01 |
|  | 5.0e−09 |  |  |
| 1 |  | 1.1e−05 | 6.0e−06 |
| 2 |  | 4.7e−01 | 7.8e−02 |
|  | 4.7e−10 |  |  |
| 1 |  | 1.0e−06 | 6.5e−08 |
| 2 |  | 4.5e−02 | 2.7e−02 |
|  | 5.5e−11 |  |  |
| 1 |  | 1.2e−07 | 1.5e−08 |
| 2 |  | 5.2e−03 | 2.5e−03 |

Figure 9.1. *The $\kappa^2$ effect.*

31.  From the backward error bound (9.3) for orthogonal triangularization
we find that solutions of the least squares problem computed by orthogonal
triangularization satisfy

$$\frac{\|\tilde{b} - b\|}{\|b\|} \lesssim c_{nk} \left( \kappa(X) + \kappa^2(X) \frac{\|r\|}{\|X\|\|b\|} \right) \epsilon_{\mathrm{M}}. \tag{9.13}$$

On the other hand solutions computed by the normal equations satisfy

$$\frac{\|\tilde{b} - b\|}{\|b\|} \lesssim d_{nk} \kappa(A) \left( 1 + \frac{\|y\|}{\|X\|\|\tilde{b}\|} \right) \epsilon_{\mathrm{M}}.$$

Now $\|A\| = \|X^{\mathrm{T}}X\| \leq \|X\|^2$ and $\|A^{-1}\| = \|X^{\dagger}(X^{\dagger})^{\mathrm{T}}\| \leq \|X^{\dagger}\|^2$. Hence

$$\kappa(A) \leq \kappa^2(X)$$

(with equality for the spectral norm). If we incorporate the term $\|y\|/(\|X\|\|\tilde{b}\|)$

into $d_{kn}$, we get the bound

$$\frac{\|\tilde{b} - b\|}{\|b\|} \lesssim d_{nk}\kappa^2(X)\epsilon_{\mathrm{M}}. \tag{9.14}$$

32. Comparing (9.13) and (9.14), we see that in general we can expect orthogonal triangularization to produce more accurate solutions than the normal equations. However, because the normal equations are computed by a simpler algorithm, the constant $d_{nk}$ will in general be smaller than $c_{nk}$. Hence if the residual is small or $\kappa(X)$ is near one, the normal equations will compute a slightly more accurate solution.

## Summary

33. So which method should one use to use to solve least squares problems? Actually, this is not the right question. What we have are two numerical tools, and we should be asking how should they be used. Here are some observations.

It is hard to argue against the normal equations in double precision. Very few problems are so ill conditioned that the results will be too inaccurate to use. The method is easier to code and faster, especially when $X$ has many zero elements. Because the normal equations have been around for almost two centuries, many auxiliary procedures have been cast in terms of them. Although there are **QR** counterparts for these procedures, they are not always immediately obvious. For this reason statisticians tend to prefer the normal equations, and there is no reason why they should not.

On the other hand, if one wants a general-purpose algorithm to run at all levels of precision, orthogonal triangularization is the winner because of its stability. Unless the data is more accurate than the precision of the computation, the user gets an answer that is no more inaccurate than he or she deserves.

- Linear and Cubic Splines

# Linear and Cubic Splines

Piecewise Linear Interpolation
The Error in $L(f)$
Approximation in the $\infty$-Norm
Hat Functions
Integration
Least Squares Approximation
Implementation Issues

## Piecewise linear interpolation

1. Depending on who you talk to, a spline is a function defined on a sequence of intervals by a corresponding sequence of polynomials. At the points where the intervals touch — that is, where the polynomials change — the function and some of its derivatives are required to be continuous. The higher the degree of the polynomials, the more continuity we can demand. We will start with linear polynomials.[11]

2. The ingredients for a linear (interpolating) spline are the following.

    1. A set of $n+1$ points $t_0 < t_1 < \cdots < t_n$ on the real line. These points are called *knots*.

    2. A set of ordinates $f_0, f_1, \ldots, f_n$. These may be values of a function $f$ at the points $t_i$, but it is not necessary to assume so.

What we want to find is a function $\ell(t)$ that satisfies the following conditions.

    1. $\ell(t)$ is a linear function in the intervals $[t_i, t_{i+1}]$.
    2. $\ell$ is continuous in $[t_0, t_n]$.               (10.1)
    3. $\ell(t_i) = f_i \ (i = 0, 1, \ldots, n)$.

The first two conditions are what makes our function a spline — we call it a *linear spline*. The third condition pins down the spline — it must interpolate the values $f_i$.

3. It is easy to show that a function satisfying (10.1) exists. Define $\ell_i$ by

$$\ell_i(t) = f_i + \frac{f_{i+1} - f_i}{t_{i+1} - t_i}(t - t_i).$$

---

[11] I am indebted to Carl de Boor's excellent book, *A Practical Guide to Splines*, for much of the material in this lecture.

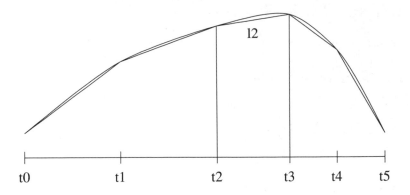

Figure 10.1. *A linear spline.*

Then $\ell_i$ is clearly linear, and it satisfies $\ell_i(t_i) = f_i$ and $\ell_i(t_{i+1}) = f_{i+1}$. Moreover, if we define $\ell$ on $[t_0, t_n]$ by

$$\ell(t) = \ell_i(t), \qquad t \in [t_i, t_{i+1}],$$

then $\ell$ is piecewise linear, interpolates the $f_i$, and is continuous [because
$\ell_{i-1}(t_i) = f_i = \ell_i(t_i)$].[12] The situation is illustrated in Figure 10.1. The
piecewise linear curve is $\ell$. The smooth curve is a function $f$ interpolated by
$\ell$ at the knots.

4. The spline $\ell$ is unique, since each function $\ell_i$ is uniquely determined by the
fact that it is linear and that $\ell_i(t_i) = f_i$ and $\ell_i(t_{i+1}) = f_{i+1}$.

5. We have solved the problem of determining $\ell$ by breaking it up into a
set of individual linear interpolation problems. But our focus will be on the
function $\ell$ in its entirety. For example, given a set of knots $t_0 < \cdots < t_n$, we
can define the space $\mathcal{L}$ of all linear splines over those knots. Note that this is
a linear space — linear combinations of continuous, piecewise linear functions
are continuous and piecewise linear.

6. We can also define an operator $L(f)$ that takes a function $f$ and produces
the linear spline $\ell$ that interpolates $f$ at the knots. Note that if $\ell \in \mathcal{L}$, then
$\ell$ interpolates itself at the knots. Since the interpolant is unique we have the
following result.

| If $\ell \in \mathcal{L}$, then $L(\ell) = \ell$. |
|---|

[12]If you are alert, you will have noticed that the interpolation property insures continuity; i.e., the second condition in (10.1) is redundant. However, in some generalizations the
condition is necessary — e.g., when the interpolation points and the knots do not coincide.

In particular, for any $f$

$$L[L(f)] = L(f). \tag{10.2}$$

Regarding $L$ as an operator, we have $L^2 = L$. Thus, $L$ can be though of as a projection onto $\mathcal{L}$ — it maps onto $\mathcal{L}$, and it leaves members of $\mathcal{L}$ unaltered.

## The error in $L(f)$

7. We now turn to the question of how well $\ell = L(f)$ approximates $f$. We shall assume that $f$ is twice continuously differentiable. It then follows from the error formula for polynomial interpolation that whenever $t \in [t_i, t_{i+1}]$

$$f(t) - \ell_i(t) = \frac{f''(\xi_t)}{2}(t - t_i)(t - t_{i+1}) \quad \text{for some} \quad \xi_t \in [t_i, t_{i+1}].$$

To convert this formula into a usable bound, let

$$h_i = t_{i+1} - t_i, \qquad i = 0, 1, \ldots, n - 1,$$

be the *mesh size at $t_i$*. Then for $t \in [t_i, t_{i+1}]$ we have $|(t - t_i)(t - t_{i+1})| \le h_i^2/4$. Hence if

$$|f''(t)| \le M_i, \qquad t \in [t_i, t_{i+1}],$$

then

$$|f(t) - \ell_i(t)| \le \frac{M_i}{8} h_i^2, \qquad t \in [t_i, t_{i+1}]. \tag{10.3}$$

Thus we have the following result.

---

Let $|f''(t)| \le M$ for $t$ in the interval $[t_0, t_n]$, and let

$$h_{\max} = \max\{h_i : i = 0, \ldots, n - 1\}. \tag{10.4}$$

Then

$$|f(t) - L(f)(t)| \le \frac{M}{8} h_{\max}^2. \tag{10.5}$$

---

8. The bound (10.5) shows that the spline approximation converges as the mesh size goes to zero. The rate of convergence is proportional to the square of $h_{\max}$. Although this convergence is not very rapid as such things go, it is fast enough to make it difficult to distinguish $f$ from $L(f)$ in parts of Figure 10.1.

9. If we are lucky enough to be able to choose our knots, the local bound (10.3) provides some guidance on how to proceed. We would like to keep the error roughly equal in each interval between the knots. Since the error bound is proportional to $M_i h_i^2$, where $M_i$ represents the second derivative of $f$ in the interval, we should choose the knot spacing so that the $h_i$ are inversely proportional to $\sqrt{M_i}$. In other words, the knots should be densest where the curvature of $f$ is greatest. We see this in Figure 10.1, where it is clear that some extra knots between $t_2$ and $t_5$ would reduce the error.

## Approximation in the ∞-norm

10. The bound (10.4) can be written in the form

$$\|f - \ell\|_\infty \leq \frac{M}{8} h_{\max}^2,$$

where $\| \cdot \|_\infty$ is the usual $\infty$-norm on $[t_0, t_n]$ (see §1.14). This suggests that we could replace the third condition in (10.1) with

$$3'. \quad \|f - \ell\|_\infty = \min.$$

If such a spline exists, it will certainly reduce the error. But we are going to show that it can't reduce the error by much.

11. Let $\ell = L(f)$. Then because each $\ell_i$ attains its greatest absolute value at an endpoint of its interval, we have

$$\|L(f)\|_\infty = \max_i |\ell(t_i)| = \max_i |f(t_i)| \leq \|f\|_\infty. \qquad (10.6)$$

Now for any $\hat{\ell} \in \mathcal{L}$ we have

$$
\begin{aligned}
\|f - L(f)\|_\infty &= \|(f - \hat{\ell}) + (\hat{\ell} - L(f))\|_\infty \\
&= \|(f - \hat{\ell}) + L(\hat{\ell} - f)\|_\infty \qquad \text{by (10.2)} \\
&\leq \|(f - \hat{\ell})\|_\infty + \|L(\hat{\ell} - f)\|_\infty \\
&\leq 2\|(f - \hat{\ell})\|_\infty \qquad\qquad \text{by (10.6).}
\end{aligned}
$$

Thus the $\infty$-norm of the error in $L(f)$ is no more than twice the error in *any* other approximation in $\mathcal{L}$. Passing to an optimal spline can reduce the error by at most a factor of two.

## Hat functions

12. Before we go on to other uses of the linear spline, it will be convenient to introduce a basis for $\mathcal{L}$ — the basis of hat functions. Specifically, the $i$th hat function $c_i$ is the unique member of $\mathcal{L}$ defined by

$$
c_i(t_j) = \begin{cases} 1 & \text{if } i = j, \\ 0 & \text{if } i \neq j. \end{cases}
$$

The name hat function comes about because under some circumstances the graph of the function resembles a triangular hat, as pictured below.[13]

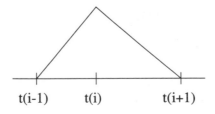

t(i-1)          t(i)                t(i+1)

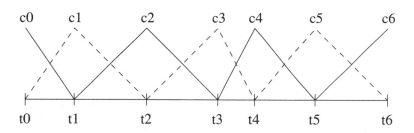

Figure 10.2. *Hat functions on a mesh.*

Figure 10.2 shows a set of hat functions over a mesh of seven knots. To distinguish them I have represented them alternately by solid and dashed lines.

13. It is easy to write down equations for the hat functions. Specifically,

$$c_i(t) = \begin{cases} (t - t_{i-1})/h_{i-1} & \text{if } t \in [t_{i-1}, t_i], \\ (t_{i+1} - t)/h_i & \text{if } t \in [t_i, t_{i+1}], \\ 0 & \text{elsewhere.} \end{cases}$$

The hat function $c_0$ omits the first of the above cases; $c_n$, the second.

14. In terms of hat functions, the linear spline interpolant has the following form:

$$L(f) = \sum_{i=0}^{n} f_i c_i. \tag{10.7}$$

In fact the sum on the right, being a linear combination of members of $\mathcal{L}$, is itself a member of $\mathcal{L}$. Moreover,

$$L(f)(t_j) = \sum_{i=0}^{n} f_i c_i(t_j) = f_j c_j(t_j) = f_j,$$

so that the spline interpolates $f$. It is worth noting that the hat functions are analogous to the Lagrange basis for polynomial interpolation.

## Integration

15. Since $L(f)$ approximates $f$ on the interval $[t_0, t_n]$ it is natural to expect its integral over the same interval to approximate the integral of $f$. From (10.7), we see that the problem of integrating $L(f)$ reduces to that of integrating the individual hat functions $c_i$.

16. These integrals can be easily computed with the aid of the following picture.

---

[13]They are also called *chapeau* functions from the French for hat.

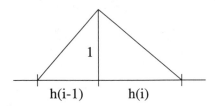

It shows the area under $c_i$ is that of a triangle of height one and base $h_{i-1}+h_i$. Since the area of a triangle is one-half its base times its height, we have

$$\int_{t_0}^{t_n} c_i(t)\,dt = \frac{h_{i-1}+h_i}{2}.$$

The hat functions $c_0$ and $c_n$ must be considered separately:

$$\int_{t_0}^{t_n} c_0(t)\,dt = \frac{h_0}{2} \quad \text{and} \quad \int_{t_0}^{t_n} c_n(t)\,dt = \frac{h_{n-1}}{2}.$$

17.  Combining these results we find that

$$T(f) \equiv \int_{t_0}^{t_n} L(f)(t)\,dt = \frac{h_0}{2} f_0 + \frac{h_0+h_1}{2} f_1 + \cdots + \frac{h_{n-1}+h_{n-1}}{2} f_{n-1} + \frac{h_{n-1}}{2} f_n.$$

If the knots are equally spaced with common spacing $h$, then

$$T(f) = h(\tfrac{1}{2}f_0 + f_1 + \cdots + f_{n-1} + \tfrac{1}{2}f_n).$$

This quadrature formula is called the *iterated trapezoidal rule* (or sometimes just the trapezoidal rule).

18.  By integrating the bound (10.5), we can obtain an error formula for $T(f)$, namely,

$$|\smallint f - T(f)| \le \frac{(t_n - t_0)M}{8} h_{\max}^2. \tag{10.8}$$

This is not the best possible bound. A more careful analysis allows us to replace the 8 in the denominator of (10.8) by 12.

## Least squares approximation

19.  As a final illustration of linear splines, we will show how they can be used to approximate a function in the least squares sense. Specifically, we can determine a member $\ell \in \mathcal{L}$ such that

$$\|f - \ell\|_2 = \min.$$

Here the 2-norm is taken in the continuous sense.

20. We will seek $\ell$ in the form

$$\ell(t) = \sum_{i=0}^{n} b_i c_i(t),$$

where the $c_i$ are the hat functions for the system of knots. The solution of this problem amounts to computing the normal equations for the $b_i$'s.

21. The $(i,j)$-element of the matrix $A$ of the normal equations is $\int c_i c_j$.[14] Now if $|i - j| > 1$, the product $c_i(t)_j(t)$ is identically zero in $[t_0, t_n]$, and hence $\alpha_{ij} = 0$. Thus our matrix has a Wilkinson diagram of the form

$$\begin{pmatrix} X & X & & & & & \\ X & X & X & & & & \\ & X & X & X & & & \\ & & \ddots & \ddots & \ddots & & \\ & & & X & X & X \\ & & & & X & X \end{pmatrix}.$$

Such a matrix is said to be *tridiagonal*.

The actual integrals are trivial though tedious to evaluate, and we will just write down the matrix of the normal equations.

$$A = \begin{pmatrix} \frac{h_0}{3} & \frac{h_0}{6} & & & & \\ \frac{h_0}{6} & \frac{h_0+h_1}{3} & \frac{h_1}{6} & & & \\ & \frac{h_2}{6} & \frac{h_1+h_2}{3} & \frac{h_2}{6} & & \\ & & \ddots & \ddots & \ddots & \\ & & & \frac{h_{n-2}}{6} & \frac{h_{n-2}+h_{n-1}}{3} & \frac{h_{n-1}}{6} \\ & & & & \frac{h_{n-1}}{6} & \frac{h_{n-1}}{3} \end{pmatrix}.$$

22. The components of the right-hand side of the normal equations are the integrals $\int c_i f$. Usually the evaluation of these integrals will be the hardest part of the procedure. However, we are helped by the fact that only integrals of $f(t)$ and $tf(t)$ are required.

23. Turning to the solution of the normal equations, note that the matrix $A$ has three properties.

1. $A$ is tridiagonal.

2. $A$ is positive definite.

3. Each diagonal element of $A$ is twice the sum of the off-diagonal elements in the same row.

---

[14]In this indexing scheme, the leading element of $A$ is $\alpha_{00}$.

The first property says that we can solve the normal equations by Gaussian elimination in $O(n)$ operations. The second tells us that pivoting is unnecessary in Gaussian elimination, which allows us to reduce the order constant in the work. The third property — an example of *strict diagonal dominance* — says that the system, properly scaled, will be well conditioned, and we can expect accurate solutions to the normal equations. (We will give a systematic treatment of diagonally dominant matrices in Lecture 24.)

24. We can also fit linear splines to sets of points. Specifically, suppose we are given $m$ points $(x_k, y_k)$. Then we can attempt to determine a linear spline $\ell$ over the knots $t_i$ such that

$$\sum_k [y_k - \ell(x_k)]^2 = \min.$$

Note that the abscissas $x_k$ are not in general the same as the knots.

The procedure is much the same as in the continuous case. We expand $\ell$ in hat functions and then form the normal equations. For example, the elements of the matrix of the normal equations are

$$\alpha_{ij} = \sum_k c_i(x_k)c_j(x_k).$$

In practice, we would only sum over indices $k$ for which both $c_i(x_k)$ and $c_j(x_k)$ are nonzero.

Once again, the matrix $A$ of the normal equations is tridiagonal. If there is at least one $x_k$ between each knot, it will also be positive definite. However, we cannot guarantee that the system is well conditioned. The trouble is that if there are no $x_k$'s in the two intervals supporting one of the hat functions, the corresponding row of the normal equations will be zero. Moving a point to slightly inside the union of these intervals will make the system nonsingular but ill conditioned. Generally speaking, you should select knots so that each hat function has some points near its peak.

## Implementation issues

25. If $t \in [t_i, t_{i+1}]$, we can evaluate $L(f)$ by the formula

$$L(f)(t) = f_i + \frac{f_{i+1} - f_i}{t_{i+1} - t_i}(t - t_i).$$

The evaluation of this formula requires four additions, one division, and one multiplication.

If it is required to evaluate the spline many times over the course of a calculation, it may pay to precompute the numbers

$$d_i = \frac{f_{i+1} - f_i}{t_{i+1} - t_i}$$

and evaluate the spline in the form

$$L(f)(t) = f_i + d_i(t - t_i).$$

By doing this we trade faster evaluation of $L(f)$ against the extra memory needed to store the $d_i$. For the linear spline, this kind of economization does not make much difference; but in more complicated settings one should pay close attention to how things are evaluated.

26. We cannot evaluate a spline at a point $t$ until we find an integer $i$ such that $t \in [t_i, t_{i+1}]$. The determination of $i$ is a potential source of trouble and is worth discussing at some length.

27. When the knots are equally spaced, we can locate the interval by a simple computation. For if the common mesh size is $h$, we can write $t_i = t_0 + ih$. Now suppose that $t \in [t_i, t_{i+1})$. Then

$$t = t_0 + ih + r,$$

where $r < h$. It follows that

$$\frac{t - t_0}{h} = i + \frac{r}{h}.$$

Since $r/h < 1$, $i$ is the smallest integer less than or equal to $(t - t_0)/h$. In terms of the floor function $\lfloor \cdot \rfloor$,

$$i = \left\lfloor \frac{t - t_0}{h} \right\rfloor.$$

In most programming languages simply assigning the quantity $(t - t_0)/h$ to an integer variable $i$ evaluates the floor function.

28. If the knots are not equally spaced, we must search for the required interval. Here is a little program that will do the job.

```
1.   find(t, i)
2.       if (t < t_0 or t > t_n) error; fi
3.       i = 0
4.       while (t > t_{i+1})                                    (10.9)
5.           i = i+1
6.       end while
7.   end find
```

After making sure that $t$ is in the interval, the program searches from $i = 0$ until it finds the first $i$ for which $t \leq t_{i+1}$. This is the required index.

Unfortunately, if $t$ is near $t_n$, the program will take $O(n)$ time to find $i$. This should be compared to the $O(1)$ time for a spline evaluation. If the number of knots is large, we may spend most of our time looking for $i$.

29.   An alternative is a technique called *binary search*. It is best described by
a program.

```
 1.   find(t, i)
 2.       if (t < t₀ or t > tₙ) error; fi
 3.       i = 0; j = n
 4.       while (j−i > 1)
 5.           k = ⌊(i+j)/2⌋
 6.           if (t < tₖ)
 7.               j = k
 8.           else
 9.               i = k
10.           end if
11.       end while
12.   end find
```

At the beginning of the while loop, $t \in [t_i, t_j]$. If $j = i + 1$, we are finished.
Otherwise, an integer $k$ is determined that is roughly halfway between $i$ and $j$.
Then $i$ or $j$ is set to $k$ depending on which assignment gives an interval containing $t$. Since each iteration of the while loop reduces the distance between
$i$ and $j$ by a factor of about two, the program must terminate after roughly
$\log_2 n$ steps. This is much better than the $O(n)$ upper bound for (10.9).

30.   Sometimes a knowledge of the application will help decide on a search
program. For example, it may happen that $t$ varies slowly, in which case the
interval for the current value of $t$ will be found near the old interval. The
following program takes advantage of the fact that we have a good starting
point for a linear search. Here the input $i$ is any integer in $[0, n-1]$—though
ordinarily it would be the last value returned.

```
 1.   find(t, i)
 2.       if (t < t₀ or t > tₙ) error; fi
 3.       while (t < tᵢ)
 4.           i = i−1
 5.       end while                                                    (10.10)
 6.       while (t ≥ tᵢ₊₁)
 7.           i = i+1
 8.       end while
 9.   end find
```

This is again a linear search, as in (10.9), but launched from a point that is
presumed good. If $t$ varies slowly enough, it will beat a binary search, which
will take a full $\log_2 n$ iterations before it finds $i$.

31.   There are many embellishments that could be made on these algorithms.
For example, if $t$ is likely to remain in the same interval for several evaluations

of the spline, one could preface (10.10) with a special test for this case, which would save two comparisons. But the real lesson to be learned from all this is: Know your application and make sure your software fits.

# Linear and Cubic Splines

Cubic Splines
Derivation of the Cubic Spline
End Conditions
Convergence
Locality

## Cubic splines

1. Historically, a spline is a thin, flexible strip of wood used to draw smooth curves through a set of points or knots. The strip is pinned to the knots but otherwise allowed it to adjust itself freely. Under idealized conditions, the resulting curve is a piecewise cubic with continuous first and second derivatives. We call such a function a cubic spline.[15]

2. The cubic spline eliminates two problems with piecewise linear interpolation. First, linear splines converge rather slowly. Second, linear splines have corners at the knots and are not suitable for applications that demand smooth approximations. Just imagine a cross section of an airplane wing in the form of a linear spline and you will quickly decide to go by rail.

3. In deriving the cubic spline, we will begin as usual with a set of $n+1$ knots $t_0 < t_1 < \cdots < t_n$ with mesh sizes $h_i = t_{i+1} - t_i$. A function $g$ defined on $[t_0, t_n]$ is a *cubic spline interpolating the points* $f_0, \ldots, f_n$ if it satisfies the following conditions.

1. $g(t)$ is a cubic polynomial $p_i(t)$ in the interval $[t_i, t_{i+1}]$.
2. $p_i(t_i) = f_i$, $i = 0, \ldots, n-1$.
3. $p_i(t_{i+1}) = f_{i+1}$, $i = 0, \ldots, n-1$.
4. $p_i'(t_{i+1}) = p_{i+1}'(t_{i+1})$, $i = 0, \ldots, n-2$.
5. $p_i''(t_{i+1}) = p_{i+1}''(t_{i+1})$, $i = 0, \ldots, n-2$.

The second and third conditions say that $g$ interpolates the $f_i$ and is continuous. The fourth and fifth conditions say that $g$ has continuous first and second derivatives.

4. A natural strategy for determining a spline is to write $p_i$ in some suitable form, say

$$p_i = a_i + b_i(t - t_i) + c_i(t - t_i)^2 + d_i(t - t_i)^3$$

[15]There is a curious shift of meaning here. The word "spline" started out as a piece of wood that determines a curve and has ended up as the curve itself.

and then use the above conditions to write down linear equations for the coefficients. This approach has two disadvantages. First, it results in a system
with an unnecessarily large bandwidth whose coefficients may vary by many
orders of magnitude. Second, the system has $4n$ unknowns, but the conditions
provide only $4n - 2$ equations. This indeterminacy is, of course, intrinsic to
the problem; but the direct approach gives little insight into how to resolve it.
Accordingly, we will take another approach.

## Derivation of the cubic spline

5. We begin by assuming that we know the second derivatives $s_i = g''(t_i)$.
Then $p_i''(t_i) = s_i$ and $p_i''(t_{i+1}) = s_{i+1}$. Since $p_i$ is linear, we can write it in the
form

$$p_i''(t) = \frac{s_i}{h_i}(t_{i+1} - t) + \frac{s_{i+1}}{h_i}(t - t_i). \tag{11.1}$$

Integrating (11.1), we find that

$$p_i'(t) = b_i - \frac{s_i}{2h_i}(t_{i+1} - t)^2 + \frac{s_{i+1}}{2h_i}(t - t_i)^2, \tag{11.2}$$

where $b_i$ is a constant to be determined. Integrating again, we get

$$p_i(t) = a_i + b_i(t - t_i) + \frac{s_i}{6h_i}(t_{i+1} - t)^3 + \frac{s_{i+1}}{h_i}(t - t_i)^3,$$

where again $a_i$ is to be determined.

6. The coefficients $a_i$ and $b_i$ are easily calculated. From the condition $p_i(t_i) = f_i$, we have

$$a_i + \frac{s_i h_i^2}{6} = f_i$$

or

$$a_i = f_i - \frac{s_i h_i^2}{6}.$$

It is worth noting that $a_i$ is just the function value $f_i$ adjusted by the second
term. This term has the dimensions of a function, since it is a second derivative
multiplied by the square of a length along the $t$-axis.

From the condition $p_i(t_{i+1}) = f_{i+1}$, we find that

$$a_i + b_i h_i + \frac{s_{i+1} h_i^2}{6} = f_{i+1}$$

or after some manipulation

$$b_i = d_i - (s_{i+1} - s_i)h_i, \tag{11.3}$$

where

$$d_i = \frac{f_{i+1} - f_i}{h_i}.$$

In the interval $[t_i, t_{i+1}]$ the cubic spline has the form

$$p_i(t) = a_i + b_i(t - t_i) + \frac{s_i}{6h_i}(t_{i+1} - t)^3 + \frac{s_{i+1}}{h_i}(t - t_i)^3,$$

where

$$a_i = f_i - \frac{s_i h_i^2}{6} \quad \text{and} \quad b_i = d_i - (s_{i+1} - s_i)h_i$$

with

$$d_i = \frac{f_{i+1} - f_i}{h_i}.$$

The second derivatives $s_i$ satisfy the equations

$$\frac{h_i}{6}s_i + \frac{h_i + h_{i+1}}{3}s_{i+1} + \frac{h_{i+1}}{6}s_{i+2} = d_{i+1} - d_i, \qquad (11.4)$$
$$i = 1, \ldots, n - 2.$$

Figure 11.1. *Cubic spline summary.*

Note that $b_i$ is an approximation to the derivative of $g$ (or $f$) adjusted by a quantity of the same dimension.

7. We now turn to the computation of the $s_i$. We have already used the interpolating and continuity conditions of §11.3. Equation (11.1) already presupposes the continuity of the second derivative of $g$. All that is left to determine the $s_i$ is the continuity of the first derivative — $p_i'(t_{i+1}) = p_{i+1}'(t_{i+1})$.

From (11.2) and (11.3) we have

$$p_i'(t_{i+1}) = b_i + \frac{h_i}{2}s_i = d_i + \frac{h_i}{6}s_i + \frac{h_i}{3}s_{i+1}.$$

Similarly

$$p_{i+1}'(t_{i+1}) = b_{i+1} - \frac{h_{i+1}}{2}s_{i+2} = d_{i+1} - \frac{h_{i+1}}{6}s_{i+2} - \frac{h_{i+1}}{3}s_{i+1}.$$

Combining these two equations, we get

$$\frac{h_i}{6}s_i + \frac{h_i + h_{i+1}}{3}s_{i+1} + \frac{h_{i+1}}{6}s_{i+2} = d_{i+1} - d_i, \qquad i = 1, \ldots, n - 1.$$

## End conditions

8. The results of our calculations are summarized in Figure 11.1. It only remains to solve the equations (11.4) for the $s_i$. These equations are tridiagonal

and diagonally dominant. But there are only $n-1$ of them to determine $n+1$ second derivatives. We must find two additional conditions.

9. If we can estimate the second derivative at the ends of the mesh, we can simply set $s_0$ and $s_n$ to these values. In this case the first and last equations of the system become

$$\frac{h_0 + h_1}{3} s_1 + \frac{h_1}{6} s_2 = d_1 - d_0 - \frac{h_0}{6} s_0 \qquad (11.5)$$

and

$$\frac{h_{n-2}}{6} s_{n-2} + \frac{h_{n-2} + h_{n-1}}{3} s_{n-1} = d_{n-1} - d_{n-2} - \frac{h_{n-1}}{6} s_n. \qquad (11.6)$$

When the conditions (11.5) and (11.6) are combined with (11.4) (for $i = 2, \ldots, n-3$), we obtain a tridiagonal, diagonally dominant system for the second derivatives $s_i$ $(i = 1, \ldots, n-1)$. This system can be solved stably by Gaussian elimination without pivoting in $O(n)$ time.

10. When we take $s_0 = s_n = 0$, we get what is called a *natural spline*. In terms of mechanical curve drawing (§11.1), it corresponds to allowing the spline (i.e., the strip of wood) to emerge freely from the endpoints. In spite of the name "natural" this choice has little to recommend it. In general, the function $f$ will not have zero second derivatives at the endpoints, and imposing such a condition on $g$ will worsen the approximation — at least around the endpoints.

11. In the absence of information about the function at the endpoints, a good alternative is to require that $p_0 = p_1$ and $p_{n-2} = p_{n-1}$. Since $p_0$ and $p_1$ have the same first and second derivatives at $t_1$ and are cubic, all we need to do is to require that the third derivatives be equal. Since

$$p_i''' = \frac{s_{i+1} - s_i}{h_i},$$

we get the equation

$$h_1 s_0 + (h_0 + h_1) s_1 - h_0 s_2 = 0,$$

which can be appended to the beginning of the system (11.4). You should verify that after one step of Gaussian elimination, the system returns to a tridiagonal, diagonally dominant system. The condition at the other end is handled similarly. These conditions are sometimes called *not-a-knot conditions* because the form of the approximation does not change across $t_1$ and $t_{n-1}$.

12. If one specifies the derivatives at the endpoints, the result is called a *complete spline*. The details are left as an exercise.

## Convergence

13. If the interpolation points $f_i$ come from a function $f$, it is natural to ask if the spline converges to $f$. For a complete spline in which the derivatives at the end are taken to be the corresponding derivatives of $f$, we have the following result.

> Let $f$ have a continuous fourth derivative, and let $g$ be the complete cubic spline interpolating $f$ at the knots $t_0, \ldots, t_n$. Let $h_{\max}$ be the maximum mesh size. Then
>
> $$
> \begin{array}{ll}
> 1. & \|f - g\|_\infty \ \le \ \frac{5}{384}\|f^{(iv)}\|_\infty h_{\max}^4, \\[4pt]
> 2. & \|f' - g'\|_\infty \ \le \ \frac{1}{24}\|f^{(iv)}\|_\infty h_{\max}^3, \qquad\qquad (11.7) \\[4pt]
> 3. & \|f'' - g''\|_\infty \le \ \frac{3}{8}\|f^{(iv)}\|_\infty h_{\max}^2.
> \end{array}
> $$

14. The spline converges to $f$ as the fourth power of the maximum step size. The derivatives also converge, albeit more slowly. Although we have not given the result here, even the third derivative converges, provided the mesh sizes go to zero in a uniform manner.

## Locality

15. As nice as the bounds (11.7) are, they are misleading in that they depend on the maximum mesh size $h_{\max}$. We have already encountered this problem with the bound

$$\|f - L(f)\|_\infty \le \tfrac{1}{8}\|f''\|_\infty h_{\max}^2$$

for piecewise linear interpolation [cf. (10.5)]. Although it appears to depend on the maximum mesh size, in reality each of the linear functions $\ell_i$ of which $L(f)$ is composed has its own error bound. This is the reason we can take widely spaced knots where the second derivative of $f$ is small without compromising the accuracy of the approximation in other parts of the interval $[t_0, t_n]$. It would be a shame if we couldn't do the same with cubic splines. It turns out that we can.

16. To see what is going on, it will be convenient to start with equally spaced knots. In this case the equations (11.4) have the form

$$\frac{h}{6}s_i + \frac{2h}{3}s_{i+1} + \frac{h}{6}s_{i+2} = d_{i+1} - d_i.$$

If we divide each equation by $2h/3$, the equations become

$$\tfrac{1}{4}s_i + s_{i+1} + \tfrac{1}{4}s_{i+2} = c_i,$$

where

$$c_i = \frac{3(d_{i+1} - d_i)}{2h}.$$

| | | | | | | | | | |
|---|---|---|---|---|---|---|---|---|---|
| 1e+00 | 3e−01 | 8e−02 | 2e−02 | 6e−03 | 1e−03 | 4e−04 | 1e−04 | 3e−05 | 7e−06 |
| 3e−01 | 1e+00 | 3e−01 | 8e−02 | 2e−02 | 6e−03 | 2e−03 | 4e−04 | 1e−04 | 3e−05 |
| 8e−02 | 3e−01 | 1e+00 | 3e−01 | 8e−02 | 2e−02 | 6e−03 | 2e−03 | 4e−04 | 1e−04 |
| 2e−02 | 8e−02 | 3e−01 | 1e+00 | 3e−01 | 8e−02 | 2e−02 | 6e−03 | 2e−03 | 4e−04 |
| 6e−03 | 2e−02 | 8e−02 | 3e−01 | 1e+00 | 3e−01 | 8e−02 | 2e−02 | 6e−03 | 1e−03 |
| 1e−03 | 6e−03 | 2e−02 | 8e−02 | 3e−01 | 1e+00 | 3e−01 | 8e−02 | 2e−02 | 6e−03 |
| 4e−04 | 2e−03 | 6e−03 | 2e−02 | 8e−02 | 3e−01 | 1e+00 | 3e−01 | 8e−02 | 2e−02 |
| 1e−04 | 4e−04 | 2e−03 | 6e−03 | 2e−02 | 8e−02 | 3e−01 | 1e+00 | 3e−01 | 8e−02 |
| 3e−05 | 1e−04 | 4e−04 | 2e−03 | 6e−03 | 2e−02 | 8e−02 | 3e−01 | 1e+00 | 3e−01 |
| 7e−06 | 3e−05 | 1e−04 | 4e−04 | 1e−03 | 6e−03 | 2e−02 | 8e−02 | 3e−01 | 1e+00 |

Figure 11.2. *Transfer coefficients.*

If we specify the second derivatives at $s_0$ and $s_n$, then we may write the equations in the form

$$(I - T)s = c,$$

where

$$T = \begin{pmatrix} 0 & -\frac{1}{4} & & & \\ -\frac{1}{4} & 0 & -\frac{1}{4} & & \\ & \ddots & \ddots & \ddots & \\ & & -\frac{1}{4} & 0 & -\frac{1}{4} \\ & & & -\frac{1}{4} & 0 \end{pmatrix}.$$

Let

$$R = (I - T)^{-1}.$$

Then since $s = Rc$, we have

$$s_i = r_{i0}c_0 + r_{i1}c_1 + \cdots + r_{i,n-2}c_{n-2} + r_{i,n-1}c_{n-1}.$$

Thus a change of $\delta$ in the $j$th-element of $c$ makes a change of $r_{ij}\delta$ in $s_i$. If $r_{ij}$ is small, $s_i$ is effectively independent of $c_j$. To give them a name, let us call the $r_{ij}$ *transfer coefficients*.

Figure 11.2 exhibits the matrix of transfer coefficients for $n = 11$. Along the diagonal they are approximately one, and they drop off quickly as you move away from the diagonal. In other words, if two knots are well separated, the function values near one of them have little to do with the second derivative calculated at the other. For example, in Figure 11.2 the influence of the first knot on the last is reduced by a factor of about $10^{-5}$.

17. This local behavior of splines is a consequence of the fact that the matrix $I - T$ is diagonally dominant and tridiagonal. To see this we need two facts. The first is the expansion $(I - T)^{-1}$ in an infinite series.

> Let $\| \cdot \|$ denote a consistent matrix norm (see §9.7). If $\|T\| < 1$,
> then $I - T$ is nonsingular, and
>
> $$(I - T)^{-1} = I + T + T^2 + T^3 + \cdots. \qquad (11.8)$$

A proof of this theorem can be found in many places (you might try your hand
at it yourself). The series (11.8) is a matrix analogue of the geometric series.
It is called a *Neumann series*.

18. In our application it will be convenient to use the matrix $\infty$-norm defined
by

$$\|A\|_\infty = \max_i \sum_j |a_{ij}|.$$

This norm — which for obvious reasons is also called the *row-sum norm* — is
consistent. For our application

$$\|T\|_\infty = \frac{1}{2}.$$

Consequently, the matrix of transfer coefficients is represented by the Neumann
series (11.8).

19. The second fact concerns the product of tridiagonal matrices. Let us first
look at the product of two such matrices:

$$
\begin{pmatrix}
X & X & 0 & 0 & 0 & 0 \\
X & X & X & 0 & 0 & 0 \\
0 & X & X & X & 0 & 0 \\
0 & 0 & X & X & X & 0 \\
0 & 0 & 0 & X & X & X \\
0 & 0 & 0 & 0 & X & X
\end{pmatrix}
\begin{pmatrix}
X & X & 0 & 0 & 0 & 0 \\
X & X & X & 0 & 0 & 0 \\
0 & X & X & X & 0 & 0 \\
0 & 0 & X & X & X & 0 \\
0 & 0 & 0 & X & X & X \\
0 & 0 & 0 & 0 & X & X
\end{pmatrix}
=
\begin{pmatrix}
X & X & X & 0 & 0 & 0 \\
X & X & X & X & 0 & 0 \\
X & X & X & X & X & 0 \\
0 & X & X & X & X & X \\
0 & 0 & X & X & X & X \\
0 & 0 & 0 & X & X & X
\end{pmatrix}.
$$

As can be verified from the above Wilkinson diagrams, the product is pentadi-
agonal — it has one more super- and subdiagonal. You should convince yourself
that further multiplication of this product by a tridiagonal matrix adds two
more super- and subdiagonals. In general we have the following result.

> Let $T$ be tridiagonal. If $k < |i - j|$, then the $(i, j)$-element of $T^k$ is
> zero. Otherwise put
>
> $$\mathbf{e}_i^{\mathrm{T}} T^k \mathbf{e}_j = 0 \text{ for } k < |i - j|. \qquad (11.9)$$

20. We are now in a position to get a bound on the $(i, j)$-element of $R$. We

have

$$
\begin{aligned}
|r_{ij}| &= |\mathbf{e}_i^T(I-T)^{-1}\mathbf{e}_j| \\
&= |\mathbf{e}_i^T I \mathbf{e}_j + \mathbf{e}_i^T T \mathbf{e}_j + \mathbf{e}_i^T T^2 \mathbf{e}_j + \cdots| \quad \text{Neumann series expansion} \\
&= |\mathbf{e}_i^T T^{|i-j|}\mathbf{e}_j + \mathbf{e}_i^T T^{|i-j|+1}\mathbf{e}_j + \cdots| \quad \text{(by 11.9)} \\
&\le \|T\|_\infty^{|i-j|} + \|T\|_\infty^{|i-j|+1} + \cdots \\
&= \|T\|_\infty^{|i-j|}/(1 - \|T\|_\infty) \qquad\qquad \text{sum of geometric series.}
\end{aligned}
$$

More generally we have established the following result.

> Let $\|\cdot\|$ be a consistent matrix norm. If $T$ is tridiagonal with $\|T\| < 1$, and $R = (I-T)^{-1}$, then
> $$
> |r_{ij}| \le \frac{\|T\|^{|i-j|}}{1 - \|T\|}.
> $$

21. For our problem the bound reduces to

$$
|r_{ij}| \le 2^{|i-j|+1}. \tag{11.10}
$$

This shows formally that the function values have only local influence on the individual spline functions.

22. Whenever one uses matrix norms, there is the danger of giving too much away. For example, our bound suggests a decay rate of $\frac{1}{2}$ for our transfer coefficients. But the data in Figure 11.2 suggests that the decay rate is more like $\frac{1}{4}$. In fact, by looking at the powers of $T$ you can convince yourself that when a diagonal becomes nonzero, the elements are the appropriate power of $\frac{1}{4}$.

23. It is now time to remove the restriction that the knots be equally spaced. (After all, the whole point of this investigation is to justify the local refinement of the knots.) The first step in our analysis is to scale the system so that the diagonal coefficients are one. From (11.4) we get the scaled equation

$$
\frac{h_i}{2(h_{i+1}+h_i)}s_i + s_{i+1} + \frac{h_{i+1}}{2(h_{i+1}+h_i)}s_{i+2} = 3\frac{d_{i+1}-d_i}{h_{i+1}+h_i}.
$$

Now the sums of the off-diagonal elements in this equation are $\frac{1}{2}$, and consequently the $\infty$-norm of $T$ is $\frac{1}{2}$. Thus the transfer coefficients satisfy the same bound as for equally spaced knots.

However, we must allow for the possibility that the division by $h_i + h_{i+1}$ will magnify the right-hand side and make it unduly influential. But as we have pointed out, our $d_i$'s are difference quotients—approximations to first derivatives. By a couple of applications of the mean value theorem we can show that

$$\left| \frac{d_{i+1} - d_i}{h_{i+1} + h_i} \right| \leq \max_{t \in [t_i, t_{i+2}]} |f''(t)|.$$

Consequently, it is the size of the second derivatives that determines the size of the scaled right-hand side and hence its influence on nearby knots. When a spline screeches around a corner it disturbs the neighborhood.

24. The general result can be used to show that changes in the end conditions also damp out. The trick is to account for any changes in the equations by perturbations in the right-hand side.

- Eigensystems

# Eigensystems

A System of Differential Equations
Complex Vectors and Matrices
Eigenvalues and Eigenvectors
Existence and Uniqueness
Left Eigenvectors
Real Matrices
Multiplicity and Defective Matrices
Functions of Matrices
Similarity Transformations and Diagonalization
The Schur Decomposition

## A system of differential equations

1. In the next several lectures we will be concerned with solutions of the equation

$$Ax = \lambda x,$$

where $A$ is a matrix of order $n$ and $x$ is nonzero. Since it is not immediately clear why such solutions are important, we will begin with an example.

2. In many applications it is necessary to solve the system of differential equations

$$y'(t) = Ay(t) \tag{12.1}$$

subject to the initial condition

$$y(0) = y_0. \tag{12.2}$$

Here $y(t) \in \mathbf{R}^n$ is a vector-valued function of the scalar $t$. The notation $y'(t)$ denotes componentwise differentiation:

$$y'(t) \stackrel{\text{def}}{=} (\eta_1'(t) \ \eta_2'(t) \ \cdots \ \eta_n'(t))^{\mathrm{T}}.$$

3. If we have a nontrivial solution of the equation $Ax = \lambda x$, we can write down a solution of (12.1). Specifically, let

$$y(t) = xe^{\lambda t}.$$

Then clearly $y'(t) = \lambda y(t)$. But

$$Ay(t) = Axe^{\lambda t} = \lambda xe^{\lambda t} = \lambda y(t) = y'(t).$$

105

Hence $xe^{\lambda t}$ satisfies (12.1).

4. In general the particular solution $y(t) = xe^{\lambda t}$ will not satisfy the initial condition $y(0) = y_0$. However, suppose we can find $n$ linearly independent vectors $x_j$ that satisfy

$$Ax_j = \lambda_j x_j$$

for some scalars $\lambda_j$. Since the $x_j$ form a basis for $\mathbf{R}^n$, the vector $y_0$ can be expressed uniquely in the form

$$y_0 = \gamma_1 x_1 + \gamma_2 x_2 + \cdots + \gamma_n x_n.$$

If we set

$$y(t) = \gamma_1 x_1 e^{\lambda_1 t} + \gamma_2 x_2 e^{\lambda_2 t} + \cdots + \gamma_n x_n e^{\lambda_n t},$$

then $y(0)$ is clearly $y_0$. As above, we can show that $y(t)$ satisfies (12.1). Hence $y(0)$ solves the initial value problem (12.1) and (12.2).

5. It should be stressed that we have done more than just solve an initial value problem. We have expressed the solution in such a way that we can read off many of its properties. If, for example, all the $\lambda_j$ are negative, the solution decays to zero. If one of the $\lambda_j$ is positive, the solution blows up. If a $\lambda_j$ is imaginary, say $\lambda_j = \alpha i$, then the solution has an oscillatory component, since $e^{\alpha i t} = \cos \alpha t + i \sin \alpha t$.[16]

## Complex vectors and matrices

6.   The possibility of complex numbers among the $\lambda_j$ in the above example forces us to come to grips with complex vectors and matrices. We will let $\mathbf{C}$, $\mathbf{C}^n$, and $\mathbf{C}^{m \times n}$ denote the sets of complex scalars, complex $n$-vectors, and complex $m \times n$ matrices. If addition and multiplication are defined in the usual way, almost everything about real vectors and matrices generalizes directly to their complex counterparts.

7. An important exception is that the formula

$$\|x\|_2^2 = x^{\mathrm{T}} x = \sum_j x_j^2 \qquad (12.3)$$

no longer defines a norm. The reason is that $x_j^2$ need not be positive. For example, if $x_j = i$, then its square is $-1$. Thus a "norm" defined by the inner product (12.3) could be zero or even complex.

8. The cure for this problem is to redefine the transpose operation in such a way that the inner product becomes a definite function. The *conjugate* of a complex number $\zeta = \xi + \eta i$ is the number $\bar{\zeta} = \xi - \eta i$. It is easily verified that

$$\zeta \bar{\zeta} = \xi^2 + \eta^2 \stackrel{\text{def}}{=} |\zeta|^2.$$

---

[16]This does not imply that the solution is complex. It turns out that $-\alpha i$ is also among the $\lambda_j$ and the net contribution of both is a real solution.

The conjugate $\bar{A}$ of a matrix $A$ is the matrix whose elements are $\bar{\alpha}_{ij}$.

The *conjugate transpose* of a matrix $A$ is the matrix

$$A^{\mathrm{H}} = \bar{A}^{\mathrm{T}}.$$

(The superscript H is in honor of the French mathematician Hermite.) If we define the inner product of two complex vectors $x$ and $y$ to be $x^{\mathrm{H}}y$, then

$$x^{\mathrm{H}}x = \sum_j |x_j|^2 \stackrel{\mathrm{def}}{=} \|x\|_2^2$$

becomes a norm — the natural generalization of the 2-norm to complex matrices. Since $x^{\mathrm{H}}y$ is a true inner product, the notion of orthogonality and all its consequences hold for complex vectors and matrices. (See §5.16 ff.)

9. The passage to complex matrices also involves a change of terminology. A matrix is said to be *Hermitian* if

$$A = A^{\mathrm{H}}.$$

A matrix $U$ is *unitary* if it is square and

$$U^{\mathrm{H}}U = I.$$

Hermitian and unitary matrices are the natural generalizations of symmetric and orthogonal matrices — natural in the sense that they share the essential properties of their real counterparts.

## Eigenvalues and eigenvectors

10. Let $A \in \mathbf{C}^{n \times n}$. The pair $(\lambda, x)$ is an *eigenpair* of $A$ if

$$x \neq 0 \quad \text{and} \quad Ax = \lambda x.$$

The scalar $\lambda$ is called an *eigenvalue* of $A$ and the vector $x$ is called an *eigenvector*.

11. The word eigenvalue is a hybrid translation of the German word *Eigenwert*, which was introduced by the mathematician David Hilbert. The component *eigen* is cognate with our English word "own" and means something like "particular to" or "characteristic." At one time mathematical pedants objected to the marriage of the German *eigen* to the English "value," and — as pedants will — they made matters worse by coining a gaggle of substitutes. Thus in the older literature you will encounter "characteristic value," "proper value," and "latent root" — all meaning eigenvalue.

Today the terms "eigenvalue" and "eigenvector" are the norm. In fact "eigen" has become a living English prefix, meaning having to do with eigenvalues and eigenvectors. Thus we have eigenpairs, eigensystems, and even eigencalculations. When I was a graduate student working at Oak Ridge, my office was dubbed the Eigencave after Batman's secret retreat.

## Existence and uniqueness

12.  Given their utility in solving differential equations, it would be a shame if eigenpairs did not exist. To show that they do, we rewrite the equation $Ax = \lambda x$ in the form

$$0 = \lambda x - Ax = (\lambda I - A)x.$$

Since $x \neq 0$, this implies that $\lambda I - A$ is singular. Conversely, if $\lambda I - A$ is singular, there is a nonzero vector $x$ such that $(\lambda I - A)x = 0$. Thus the eigenvalues of $A$ are precisely those values that make $\lambda I - A$ singular.

13.  A necessary and sufficient condition for a matrix to be singular is that its determinant be zero. Thus the eigenvalues of a matrix satisfy the *characteristic equation*

$$p(\lambda) \equiv \det(\lambda I - A) = 0.$$

The function $p(\lambda)$ is easily seen to be a polynomial whose highest-order term is $\lambda^n$. It is called the *characteristic polynomial*. From the theory of polynomials, we know that $p(\lambda)$ has a unique factorization in the form

$$p(\lambda) = (\lambda - \lambda_1)(\lambda - \lambda_2) \cdots (\lambda - \lambda_n). \tag{12.4}$$

If we count each $\lambda_i$ by the number of times it appears in the factorization (12.4) we see that:

> A matrix of order $n$ has $n$ eigenvalues counting multiplicities in the characteristic polynomial. The eigenvalues and their multiplicities are unique.

## Left eigenvectors

14.  If a matrix $A$ is singular, so is $A^H$. In particular if $\lambda I - A$ is singular, then $\bar{\lambda} I - A^H$ is singular. If follows that the eigenvalues of $A^H$ are the conjugates of the eigenvalues of $A$.

15.  If $\bar{\lambda}$ is an eigenvalue of $A^H$, it has an eigenvector $y$ satisfying $A^H y = \bar{\lambda} y$. Written in terms of the original matrix $A$,

$$y^H A = \lambda y^H.$$

The vector $y^H$ is called a *left eigenvector* corresponding to $\lambda$.

## Real matrices

16.  If the elements of $A$ are real, the coefficients of its characteristic polynomial are real. Now the complex zeros of a real polynomial occur in conjugate pairs. Moreover, on taking complex conjugates in the equation $Ax = \lambda x$ we get $A\bar{x} = \bar{\lambda}\bar{x}$. Hence:

> If $A$ is real, its complex eigenvalues occur in conjugate pairs. If $(\lambda, x)$ is a complex eigenpair of $A$, then $(\bar{\lambda}, \bar{x})$ is also an eigenpair.

## Multiplicity and defective matrices

17. The multiplicity of an eigenvalue as a root of the characteristic equation is called its *algebraic multiplicity*. The very fact that we qualify the word "multiplicity" suggests that there is another definition of multiplicity waiting in the wings. It is called geometric multiplicity.

18. The notion of geometric multiplicity is a consequence of the following observations. Let $x$ and $y$ be eigenvectors of $A$ corresponding to the same eigenvalue $\lambda$. Then for any scalars $\alpha$ and $\beta$

$$A(\alpha x + \beta y) = \alpha A x + \beta A y = \lambda(\alpha x + \beta y).$$

Thus if the linear combination $\alpha x + \beta y$ is nonzero, it is also an eigenvector of $A$ corresponding to $\lambda$. More generally, any nonzero linear combination of eigenvectors corresponding to $\lambda$ is also an eigenvector corresponding to $\lambda$. It follows that:

> The eigenvectors corresponding to an eigenvalue $\lambda$ of $A$ along with the zero vector form a subspace. We call the dimension of that subspace the *geometric multiplicity* of $\lambda$.

19. We shall see later that the algebraic multiplicity of an eigenvalue is always greater than or equal to its geometric multiplicity. Unfortunately, inequality can hold, as the following example shows.
    Let

$$A = \begin{pmatrix} 1 & 1 \\ 0 & 1 \end{pmatrix}. \tag{12.5}$$

The characteristic polynomial of $A$ is easily seen to be $(\lambda - 1)^2$, so that one is an eigenvalue of algebraic multiplicity two. But if we attempt to find an eigenvector from the equation

$$\begin{pmatrix} 1 & 1 \\ 0 & 1 \end{pmatrix} \begin{pmatrix} \alpha \\ \beta \end{pmatrix} = \begin{pmatrix} \alpha \\ \beta \end{pmatrix},$$

we find that $\alpha + \beta = \alpha$ and hence that $\beta = 0$. Thus all eigenvectors have the form

$$\alpha \begin{pmatrix} 1 \\ 0 \end{pmatrix}, \qquad \alpha \neq 0,$$

and the dimension of the corresponding subspace is one.

20. An eigenvalue whose algebraic multiplicity exceeds its geometric multiplicity is called *defective*. A matrix containing defective eigenvalues is also called *defective*. Defective matrices are not nice things. For example, the approach to solving systems of differential equations outlined at the beginning of this

lecture (§§12.2–12.4) does not work for defective matrices, since a defective matrix does not have $n$ linearly independent eigenvectors.[17]

21.   The eigenvalues of defective matrices are very sensitive to perturbations in the elements of the matrix. Consider, for example, the following perturbation of the matrix of (12.5):

$$\tilde{A} = \begin{pmatrix} 1 & 1 \\ \epsilon & 1 \end{pmatrix}.$$

The characteristic equation of $\tilde{A}$ is $(\lambda - 1)^2 = \epsilon$, so that the eigenvalues are

$$1 \pm \sqrt{\epsilon}.$$

Thus a perturbation of, say, $10^{-10}$ in $A$ introduces a perturbation of $10^{-5}$ in its eigenvalues!

Sensitivity is not the only ill consequence of defectiveness. Defectiveness also slows down the convergence of algorithms for finding eigenvalues.

## Functions of matrices

22.     Many computational algorithms for eigensystems transform the matrix so that the eigenvalues are more strategically placed. The simplest such transformation is a *shift of origin*.

Suppose that $Ax = \lambda x$. Then $(A - \mu I)x = (\lambda - \mu)x$. Thus subtracting a number $\mu$ from the diagonal of a matrix shifts the eigenvalues by $\mu$. The eigenvectors remain unchanged.

23.  From the equation $Ax = \lambda x$ it follows that

$$A^2 x = A(Ax) = \lambda Ax = \lambda^2 x.$$

Hence squaring a matrix squares its eigenvalues. In general, the eigenvalues of $A^k$ are the $k$th powers of the eigenvalues of $A$.

24.    If $f(t) = \gamma_0 + \gamma_1 t + \cdots + \gamma_n t^n$ is a polynomial, we can define

$$f(A) = \gamma_0 I + \gamma_1 A + \cdots + \gamma_n A^n.$$

On multiplying $f(A)$ and $x$ and simplifying, we get

$$Ax = \lambda x \implies f(A)x = f(\lambda)x.$$

Thus if we form a polynomial in $A$, the eigenvalues are transformed by the same polynomial.

---

[17] Defective eigenvalues are associated with special solutions like $te^{\lambda t}, t^2 e^{\lambda t}, \ldots$.

25. Now suppose $A$ is nonsingular, and $Ax = \lambda x$. Then $\lambda$ must be nonzero; for otherwise $x$ would be a null vector of $A$. It then follows that

$$A^{-1}x = \lambda^{-1}x.$$

Thus, inverting a matrix inverts its eigenvalues.

26. Finally let $h(t) = f(t)/g(t)$ be a rational function, and suppose that $g(A)$ is nonsingular. Then we may define $h(A)$ by

$$h(A) = g(A)^{-1}f(A).$$

It follows from the results derived above that if $Ax = \lambda x$ then

$$h(A)x = h(\lambda)x.$$

Hence the eigenvalues of $A$ are transformed by $h$.

27. The general principle is that taking a function of a matrix transforms the eigenvalues by that function and leaves the eigenvectors unchanged. This applies not only to rational functions but functions defined by a power series (we will not use this fact in the sequel). For example, if $\lambda$ is an eigenvalue of $A$, then $e^\lambda$ is an eigenvalue of

$$e^A \stackrel{\mathrm{def}}{=} \sum_{k=0}^{\infty} \frac{A^k}{k!}.$$

## Similarity transformations and diagonalization

28. A standard technique of matrix computations is to transform a matrix to one of simpler form — e.g., upper triangular form — that makes the solution of the problem at hand easy. In doing so it is important to use transformations that preserve the things you want to compute. For the eigenvalue problem the appropriate transformations are called similarity transformations.

29. Let $X$ be nonsingular. Then the matrix $X^{-1}AX$ is said to be *similar* to $A$, and the transformation $A \rightarrow X^{-1}AX$ is called a *similarity transformation.* Similarity transformations preserve the characteristic polynomial. In fact,

$$\begin{aligned} \det(\lambda I - X^{-1}AX) &= \det[X^{-1}(\lambda I - A)X] \\ &= \det(X^{-1})\det(\lambda I - A)\det(X) \\ &= \det(\lambda I - A). \end{aligned}$$

It follows that similarity transformations preserve the eigenvalues of $A$ along with their multiplicities.

A similarity transformation does not preserve eigenvectors, but it transforms them in a predictable way. Let $Ax = \lambda x$ and $y = X^{-1}x$. Then

$$(X^{-1}AX)y = \lambda y.$$

Thus, the similarity transformation $X^{-1}AX$ transforms the eigenvector $x$ to $X^{-1}x$.

30.   In trying to simplify an eigenvalue problem by similarity transformations, it is natural to shoot for a diagonal matrix. Unfortunately we cannot always achieve such a form. To see this, suppose that

$$X^{-1}AX = \Lambda = \text{diag}(\lambda_1, \ldots, \lambda_n).$$

Partitioning $X$ by columns, we have

$$A(x_1, \ldots, x_n) = (x_1, \ldots, x_n)\text{diag}(\lambda_1, \ldots, \lambda_n),$$

whence

$$Ax_i = \lambda_i x_i, \qquad i = 1, \ldots, n.$$

Thus the $n$ columns of $X$ are eigenvectors of $A$. Since they are linearly independent, the matrix $A$ cannot be defective. In other words, not all matrices can be diagonalized by a similarity. We must lower our sights.

## The Schur decomposition

31.   A natural class of matrices to use for similarity transformations are the unitary matrices. Recall that a matrix $U$ is unitary if $U^H U = I$; that is, its conjugate transpose is its inverse. It follows that if $U$ is unitary, then the transformation $U^H AU$ is a similarity transformation. It is a great advantage that we do not have to form an inverse matrix to effect the transformation.

   The orthonormality of the columns of a unitary matrix restricts the degrees of freedom we have to reduce a matrix $A$. In particular, there are just enough degrees of freedom to eliminate all the elements below the diagonal of $A$, i.e., to reduce $A$ to triangular form. A famous result of Schur says that such a reduction can always be effected.

---

Let $A \in \mathbf{C}^{n \times n}$. Then there is a unitary matrix $U$ such that

$$U^T AU = T,$$

where $T$ is upper triangular. The diagonal elements of $T$ are the eigenvalues of $A$, which can be made to appear in any order.

---

32. To establish this result, let an ordering for the eigenvalues of $A$ be given, and let $\lambda_1$ be the first eigenvalue in that ordering. Let $v_1$ be an eigenvector corresponding to $\lambda_1$ normalized so that $\|v_1\|_2 = 1$. Let the matrix $(v_1\ V_2)$ be unitary. (This matrix can be constructed by computing a QR decomposition of the vector $v_1$. See §13.2.)

Consider the matrix

$$(v_1\ V_2)^{\mathrm{H}} A (v_1\ V_2) = \begin{pmatrix} v_1^{\mathrm{H}} A v_1 & v_1^{\mathrm{H}} A V_2 \\ V_2^{\mathrm{H}} A v_1 & V_2^{\mathrm{H}} A V_2 \end{pmatrix} = \begin{pmatrix} \lambda_1 v_1^{\mathrm{T}} v_1 & h^{\mathrm{H}} \\ \lambda_1 V_2^{\mathrm{H}} v_1 & B \end{pmatrix},$$

where $h^{\mathrm{H}} = v_1^{\mathrm{H}} A V_2$ and $B = V_2^{\mathrm{H}} A V_2$. By the orthonormality of the columns of $(v_1\ V_2)$, we have $v_1^{\mathrm{H}} v_1 = 1$ and $V_2^{\mathrm{H}} v_1 = 0$. Hence

$$(v_1\ V_2)^{\mathrm{H}} A (v_1\ V_2) = \begin{pmatrix} \lambda_1 & h^{\mathrm{H}} \\ 0 & B \end{pmatrix}.$$

Now

$$\det\left[ \lambda I - \begin{pmatrix} \lambda_1 & h^{\mathrm{H}} \\ 0 & B \end{pmatrix} \right] = (\lambda - \lambda_1)\det(\lambda I - B).$$

Hence the eigenvalues of $B$ are the eigenvalues of $A$ excluding $\lambda_1$. By an obvious induction, we can find a unitary matrix $W$ such that $S = W^{\mathrm{H}} B W$ is upper triangular with the eigenvalues of $A$ (excluding $\lambda_1$) appearing in the proper order. If we set

$$U = (v_1\ V_2 W),$$

it is easily verified that $U$ is unitary and

$$T \equiv U^{\mathrm{T}} A U = \begin{pmatrix} \lambda_1 & h^{\mathrm{H}} W \\ 0 & S \end{pmatrix}.$$

Clearly $T$ is an upper triangular matrix with the eigenvalues of $A$ appearing in the prescribed order on its diagonal.

# Eigensystems

Real Schur Form
Block Diagonalization
Diagonalization
Jordan Canonical Form
Hermitian Matrices
Perturbation of a Simple Eigenvalue

## Real Schur form

1.    Most routines for solving dense nonsymmetric eigenvalue problems begin by computing a Schur form. There are three reasons. First, the form itself is useful in a number of applications. Second, it can be computed stably using orthogonal transformations. Third, if eigenvectors are needed, they can be computed cheaply from the Schur form.

If the matrix in question is real, however, the Schur form may be complex. Because complex arithmetic is expensive, complex matrices are to be avoided whenever possible. Fortunately, there is a real variant of the Schur form.

---

Let $A \in \mathbf{R}^{n \times n}$. Then there is an orthogonal matrix $U$ such that $U^{\mathrm{T}} A U$ has the block triangular form

$$T = U^{\mathrm{T}} A U = \begin{pmatrix} T_{11} & T_{12} & T_{13} & \cdots & T_{1k} \\ 0 & T_{22} & T_{23} & \cdots & T_{2k} \\ 0 & 0 & T_{33} & \cdots & T_{3k} \\ \vdots & \vdots & \vdots & & \vdots \\ 0 & 0 & 0 & \cdots & T_{kk} \end{pmatrix}.$$

The diagonal blocks are either $1 \times 1$, in which case they are real eigenvalues of $T$, or $2 \times 2$, in which case they contain a pair of complex conjugate eigenvalues of $T$. The matrix $T$ is called a *real Schur form*.

---

2.    The existence of the form is established in the same way as the existence of the Schur form, except that we pick off conjugate pairs of eigenvalues together. In outline, let $(\mu + i\nu, x + iy)$ be a complex eigenpair, so that

$$A(x + iy) = (\mu + i\nu)(x + iy).$$

Then it is easily verified that $x$ and $y$ are linearly independent and

$$A(x \ \ y) = (x \ \ y) \begin{pmatrix} \mu & \nu \\ -\nu & \mu \end{pmatrix} \equiv (x \ \ y)M.$$

(You should check that the eigenvalues of $M$ are $\mu \pm i\nu$.) Let

$$\begin{pmatrix} V_1^{\mathrm T} \\ V_2^{\mathrm T} \end{pmatrix} (x \;\; y) = \begin{pmatrix} R \\ 0 \end{pmatrix}$$

be a QR decomposition of $(x \;\; y)$, so that $V_1 = (x \;\; y)R^{-1}$. Then

$$\begin{pmatrix} V_1^{\mathrm T} \\ V_2^{\mathrm T} \end{pmatrix} AV_1 = \begin{pmatrix} V_1^{\mathrm T} \\ V_2^{\mathrm T} \end{pmatrix} A(x \;\; y)R^{-1}$$

$$= \begin{pmatrix} V_1^{\mathrm T} \\ V_2^{\mathrm T} \end{pmatrix} (x \;\; y)MR^{-1}$$

$$= \begin{pmatrix} RMR^{-1} \\ 0 \end{pmatrix}.$$

Consequently,

$$\begin{pmatrix} V_1^{\mathrm T} \\ V_2^{\mathrm T} \end{pmatrix} A(V_1 \;\; V_2) = \begin{pmatrix} RMR^{-1} & V_1^{\mathrm H}AV_2 \\ 0 & V_2^{\mathrm T}AV_2 \end{pmatrix}.$$

Thus $A$ has been transformed into a real block triangular matrix with a leading $2 \times 2$ block containing a complex conjugate pair of eigenvalues. A recursive application of the procedure to $V_2^{\mathrm T}AV_2$ completes the reduction.

## Block diagonalization

3. We now return to general complex matrices. Having reduced our matrix $A$ to Schur form by unitary similarities, it is natural to ask how much further we can reduce it by general similarity transformations. The ultimate answer is the Jordan canonical form, which we will consider later. But there is an intermediate reduction to block diagonal form that is useful in theory and practice.

4. Let us suppose that we have computed the Schur form

$$T = U^{\mathrm H}AU = \begin{pmatrix} T_{11} & T_{12} & T_{13} & \cdots & T_{1k} \\ 0 & T_{22} & T_{23} & \cdots & T_{2k} \\ 0 & 0 & T_{33} & \cdots & T_{3k} \\ \vdots & \vdots & \vdots & & \vdots \\ 0 & 0 & 0 & \cdots & T_{kk} \end{pmatrix},$$

where each $T_i$ is an upper triangular matrix with a single distinct eigenvalue on its diagonal. We are going to show how to use similarity transformations to annihilate the superdiagonal blocks.

5. Let us repartition $T$ in the form

$$T = \begin{pmatrix} B & H \\ 0 & C \end{pmatrix},$$

where $B = T_{11}$. By the construction of $T$, the eigenvalue of $B$ is distinct from the eigenvalues of $C$. We seek a matrix $P$ such that

$$\begin{pmatrix} I & -P \\ 0 & I \end{pmatrix} \begin{pmatrix} B & H \\ 0 & C \end{pmatrix} \begin{pmatrix} I & P \\ 0 & I \end{pmatrix} = \begin{pmatrix} B & 0 \\ 0 & C \end{pmatrix}. \tag{13.1}$$

Note that the right-hand side of (13.1) is indeed a similarity transformation, since

$$\begin{pmatrix} I & P \\ 0 & I \end{pmatrix}^{-1} = \begin{pmatrix} I & -P \\ 0 & I \end{pmatrix}.$$

If we compute the $(2,2)$-block of the right-hand side of (13.1) and set the result to zero, we get the following equation for $P$:

$$BP - PC = -H. \tag{13.2}$$

6. Equation (13.2) is called a *Sylvester equation*. Because $B$ and $C$ are upper triangular, we can solve it explicitly. Suppose that $P$ has $\ell$ columns. Partition (13.2) in the form

$$B(p_1 \ p_2 \ p_3 \ \cdots \ p_\ell) - (p_1 \ p_2 \ p_3 \ \cdots \ p_\ell) \begin{pmatrix} c_{11} & c_{12} & c_{13} & \cdots & c_{1\ell} \\ 0 & c_{12} & c_{13} & \cdots & c_{1\ell} \\ 0 & 0 & c_{13} & \cdots & c_{1\ell} \\ \vdots & \vdots & \vdots & & \vdots \\ 0 & 0 & 0 & \cdots & c_{1\ell} \end{pmatrix} \tag{13.3}$$

$$= -(h_1 \ h_2 \ h_3 \ \cdots \ h_\ell).$$

On computing the first column of this equation, we find that

$$Bp_1 + c_{11}p_1 = -h_1$$

or

$$(B - c_{11}I)p_1 = -h_1.$$

This is an upper triangular system. It is nonsingular because $c_{11}$, which is an eigenvalue of $C$, cannot appear on the diagonal of $B$. Hence we may solve the equation by the standard back-substitution algorithm.

If we compute the second column of (13.3), we get

$$Bp_2 - c_{12}p_1 - c_{22}p_2 = -h_2.$$

Since we already know $p_1$, we can move it to the right-hand side to get

$$(B - c_{22}I)p_2 = c_{12}p_1 - h_2.$$

Again this is a nonsingular, triangular system, which can be solved for $p_2$.

The third column of (13.3) gives

$$(B - c_{33}I)p_3 = c_{13}p_1 + c_{23}p_2 - h_2.$$

Once again this system can be solved for $p_3$.

At this point the general procedure should be apparent. We summarize it in the following algorithm.

     1.   **for** $j = 1$ **to** $\ell$
     2.       Solve the system
$$(B-C[j,j]I)P[:,j] \qquad\qquad (13.4)$$
$$= P[:,1{:}j-1]*C[1{:}j-1,j]-H[:,j]$$
     3.   **end for** $j$

7. Having eliminated $H$ we may now apply the procedure recursively to reduce $C$ to block triangular form. This process gives the following decomposition.

---

Let $A$ have $k$ distinct (possibly multiple) eigenvalues $\lambda_1, \ldots, \lambda_k$. Then there is a nonsingular matrix $X$ such that

$$X^{-1}AX = \operatorname{diag}(T_{11}, T_{22}, \ldots, T_{kk}), \qquad (13.5)$$

where $T_{ii}$ $(i = 1, \ldots, k)$ is an upper triangular matrix with $\lambda_i$ on its diagonal.

---

8. The procedure we have just described is constructive and may be useful in some applications. However, its implementation is not trivial. The main problem is that when a matrix with multiple eigenvalues is placed on a computer, rounding error will generally disturb the eigenvalues so that they are no longer multiple. If we attempt to apply the above procedure to the individual eigenvalues, some of the $p$'s (they are now vectors) will be large and the final transformation $X$ will be ill conditioned. The cure is to group nearby eigenvalues together in the Schur form. Fortunately, the procedure (13.4) does not require that $B$ have only one multiple eigenvalue — just that its eigenvalues be different from those of $C$. On the other hand, it is not easy to decide what eigenvalues to bring together. Ad hoc procedures have appeared in the literature, but none that provably work.

## Diagonalization

9. We have seen (§12.30) that any matrix that can be diagonalized by a similarity transformation is nondefective. The converse follows from the block diagonal form (13.5) and the following result.

> Let $A$ be a nondefective matrix with a single (possibly multiple) eigenvalue $\lambda$. Then $A = \lambda I$.

To establish this result, suppose $A$ is of order $n$ and let $X = (x_1 \cdots x_n)$ consist of a set of linearly independent eigenvectors of $A$. Then $X^{-1}AX = \lambda I$. Hence

$$A = \lambda X I X^{-1} = \lambda I.$$

10.   It follows that if $A$ is nondefective, then the diagonal blocks in (13.5) are themselves diagonal. Hence:

> A matrix is diagonalizable if and only if it is nondefective.

In particular, any matrix with distinct eigenvalues is diagonalizable.

## Jordan canonical form

11.  Defective matrices cannot be diagonalized, but they can be reduced beyond the simple block diagonal form (13.5). Specifically, define a *Jordan block* $J_m(\lambda)$ to be an $m \times m$ matrix of the form

$$J_m(\lambda) = \begin{pmatrix} \lambda & 1 & 0 & \cdots & 0 & 0 \\ 0 & \lambda & 1 & \cdots & 0 & 0 \\ 0 & 0 & \lambda & \cdots & 0 & 0 \\ \vdots & \vdots & \vdots & & \vdots & \vdots \\ 0 & 0 & 0 & \cdots & \lambda & 1 \\ 0 & 0 & 0 & \cdots & 0 & \lambda \end{pmatrix}.$$

Note that a Jordan block is as defective as they come. It has only one eigenvector corresponding to its single eigenvalue.

Jordan blocks are the pieces from which the *Jordan canonical form* is built.

> For any matrix $A$ there is a nonsingular matrix $X$ such that
>
> $$X^{-1}AX = \mathrm{diag}[j_{m_1}(\lambda_1), j_{m_2}(\lambda_2), \ldots, j_{m_k}(\lambda_k)].$$
>
> Except for the ordering of the diagonal blocks, the right-hand side of this decomposition is unique.

12.  The Jordan canonical form is a beautiful resolution of the nasty properties of defective matrices. And it has many theoretical applications. But in computational practice, it is disappointing. In the first place, rounding error breaks it apart. In the second place, many matrices are merely near a Jordan canonical form, and even near to more than one. For these reasons disciplines — like control theory — that once used the Jordan form as a mathematical underpinning have tended to look to other decompositions, such as the Schur form.

## Hermitian matrices

13.   About the only way you can determine if a general matrix is diagonalizable is to try to compute its eigenvectors and see if you run into trouble. Some kinds of matrices, however, are known a priori to be diagonalizable. Of these none is more important than the class of Hermitian matrices (§12.9).

14.   If $A$ is Hermitian, then $a_{ii} = \bar{a}_{ii}$. Since only real numbers can be equal to their conjugates, the diagonal elements of a Hermitian matrix are real.

15. In transforming Hermitian matrices it is important to preserve the property of being Hermitian. The most natural class of transformation is the class of unitary similarity transformations, since if $A$ is Hermitian

$$(U^{\mathrm{H}}AU)^{\mathrm{H}} = U^{\mathrm{H}}AU.$$

In particular, any Schur form of $A$ is Hermitian. But if a triangular matrix is Hermitian, then it is both lower and upper triangular — i.e., it is diagonal. Moreover, the diagonal elements are their own conjugates and are therefore real. Hence we have the following important result.

---

Let $A$ be Hermitian. Then there is a unitary matrix $U$ such that

$$U^{\mathrm{H}}AU = \mathrm{diag}(\lambda_1, \ldots, \lambda_n),$$

where the $\lambda_i$ are real. Equivalently, a Hermitian matrix has real eigenvalues and their eigenvectors can be chosen to form an orthonormal basis for $\mathbf{C}^n$.

---

16.   It is worth noting that the Hermitian matrices are an example of a more general class of matrices called *normal matrices.* A matrix $A$ is normal if $AA^{\mathrm{H}} = A^{\mathrm{H}}A$. It can be shown that any Schur form of a normal matrix is diagonal. Hence normal matrices have a complete set of orthonormal eigenvectors. However, their eigenvalues may be complex.

Aside from Hermitian matrices, the only natural class of normal matrices are the unitary matrices.

## Perturbation of a simple eigenvalue

17.   Before attempting to devise an algorithm to compute a quantity, it is always a good idea to find out how sensitive the quantity is to perturbation in the data. For eigenvalues this problem takes the following form. Let

$$Ax = \lambda x$$

and let

$$\tilde{A} = A + E,$$

where $E$ is presumed small. Is there an eigenvalue $\tilde{\lambda}$ near $\lambda$, and if so can we bound $|\tilde{\lambda} - \lambda|$ in terms of $E$?

For multiple eigenvalues this question is difficult to answer. If $\lambda$ is *simple* — that is if it has algebraic multiplicity one — then it is a differentiable function of the elements of $A$, and the perturbation problem has an elegant solution. As we did with the least squares problem, we will develop a first-order perturbation expansion.

18. We begin with a result that is interesting in its own right.

> Let $\lambda$ be a simple eigenvalue of $A$ with right and left eigenvectors $x$ and $y^H$. Then
> $$y^H x \neq 0.$$

To see this, note that if we transform $A$ by a unitary similarity to $U^H A U$ then the right and left eigenvectors transform into $U^H x$ and $y^H U$. But since $U$ is unitary,
$$(y^H U)(U^H x) = y^H x.$$

Hence the original and the transformed vectors have the same inner products, and we may work with the transformed matrix. In particular, we may assume without loss of generality that $A$ is in Schur form with $\lambda$ as its $(1,1)$-element.

Partition this form as
$$A = \begin{pmatrix} \lambda & h^H \\ 0 & C \end{pmatrix}.$$

Since $\lambda$ is simple, it is not an eigenvalue of $C$.

Now the right eigenvector of $A$ is $x = \mathbf{e}_1$. If we look for a left eigenvector in the form $y^H = (\eta_1 \ y_2^H)$, then we must have
$$(\eta_1 \ y_2^H) \begin{pmatrix} \lambda & h^H \\ 0 & C \end{pmatrix} = \lambda(\eta_1 \ y_2^H).$$

Hence
$$\eta h^H + y_2^H C = \lambda y_2^H.$$

It follows that $\eta \neq 0$, for otherwise $\lambda$ would be an eigenvalue of $C$. But $y^H x = \eta$, which establishes the result.

19. Returning now to the perturbation problem, let us write the perturbed equation in the form
$$(A + E)(x + h) = (\lambda + \mu)(x + h),$$

where it is assumed that $h$ and $\mu$ go to zero along with $E$. Expanding and throwing out second-order terms we get
$$Ax + Ah + Ex \cong \lambda + \lambda h + \mu x.$$

Since $Ax = \lambda x$, we have

$$Ah + Ex \cong \lambda h + \mu x.$$

Let $y^{\mathrm{H}}$ be the left eigenvector of $A$ corresponding to $\lambda$. Then

$$y^{\mathrm{H}} Ah + y^{\mathrm{H}} Ex \cong \lambda y^{\mathrm{H}} h + \mu y^{\mathrm{H}} x,$$

or since $y^{\mathrm{H}} A = \lambda y^{\mathrm{H}}$,

$$y^{\mathrm{H}} Ex \cong \mu y^{\mathrm{H}} x.$$

Because $\lambda$ is simple $y^{\mathrm{H}} x \neq 0$. Hence

$$\mu \cong \frac{y^{\mathrm{H}} Ex}{y^{\mathrm{H}} x}. \tag{13.6}$$

20. If we take norms in (13.6), we get

$$|\lambda - \tilde{\lambda}| \lesssim \frac{\|x\|_2 \|y\|_2}{|y^{\mathrm{H}} x|} \|E\|_2.$$

Now $|y^{\mathrm{H}} x| / \|x\|_2 \|y_2\|$ is the cosine of the angle between $x$ and $y$ (see §5.17). Hence its reciprocal is the secant of this angle. Thus we may summarize our results as follows.

---

Let $\lambda$ be a simple eigenvalue of $A$ with right eigenvector $x$ and left eigenvector $y$. For all sufficiently small $E$ there is an eigenvalue $\tilde{\lambda}$ of $A + E$ that satisfies

$$\tilde{\lambda} = \lambda + \frac{y^{\mathrm{H}} Ex}{y^{\mathrm{H}} x} + O(\|E\|_2^2).$$

Moreover,

$$|\tilde{\lambda} - \lambda| \leq \sec \angle(x, y) \|E\|_2 + O(\|E\|_2^2).$$

---

21. Thus the secant of the angle between the left and right eigenvectors of an eigenvalue is a condition number for the eigenvalue. It assumes its minimum value of one when the angle is zero and becomes infinite as the angle approaches $\frac{\pi}{2}$ — that is, as the left and right vectors become orthogonal. Since the left and right eigenvectors of a Hermitian matrix are the same, the eigenvalues of Hermitian matrices are as well conditioned as possible.

A Jordan block, on the other hand, has orthogonal left and right eigenvectors, and its eigenvalue could be said to be infinitely ill conditioned. Actually, the eigenvalue is nonanalytic, and our first-order perturbation theory does not apply. However, for matrices near a Jordan block, the eigenvectors will be very nearly orthogonal, and the eigenvalues very ill conditioned. For example, the matrix

$$A = \begin{pmatrix} 0 & 1 \\ \epsilon & 0 \end{pmatrix}$$

has the eigenvalue $\sqrt{\epsilon}$ with left and right eigenvectors

$$y^{\mathrm{T}} = (\sqrt{\epsilon} \;\; 1) \quad \text{and} \quad x = \begin{pmatrix} 1 \\ \sqrt{\epsilon} \end{pmatrix},$$

which are almost orthogonal.

# Eigensystems

A Backward Perturbation Result
The Rayleigh Quotient
Powers of Matrices
The Power Method

## A backward perturbation result

1.   Let $(\mu, x)$ be an approximate eigenpair of $A$ with $\|x\|_2 = 1$. In trying to assess the quality of this pair it is natural to look at how well $\mu$ and $x$ satisfy the equation $Ax = \mu x$. Specifically, consider the residual

$$r = Ax - \mu x.$$

If $r$ is zero, then $(\mu, x)$ is an eigenpair of $A$. If $r$ is small, what can we say about the accuracy of $\mu$ and $x$?

2.   Unfortunately, we can't say much in general. Consider, for example, the perturbed Jordan block

$$A = \begin{pmatrix} 1 & 1 \\ \epsilon & 1 \end{pmatrix},$$

where $\epsilon$ is small. If $x = (1 \ 0)^{\mathrm{T}}$ and $\mu = 1$, then

$$\|r\|_2 = \left\| \begin{pmatrix} 0 \\ \epsilon \end{pmatrix} \right\|_2 = \epsilon.$$

But as we have seen (§12.21) the eigenvalues of $A$ are $1 \pm \sqrt{\epsilon}$. Thus if $\epsilon = 10^{-10}$, the norm of the residual $r$ underestimates the perturbation in the eigenvalues by five orders of magnitude.

3. Nonetheless, the residual contains useful information about the eigensystem, as the following result shows.

For any $x \neq 0$ and any scalar $\mu$ let

$$r = Ax - \mu x.$$

Then there is a matrix

$$E = -\frac{rx^{\mathrm{H}}}{\|x\|_2^2}$$

with

$$\|E\|_p = \frac{\|r\|_2}{\|x\|_2}, \qquad p = 2, \mathrm{F},$$

such that

$$(A + E)x = \mu x.$$

The proof is a straightforward verification.

4. This *backward perturbation bound* says that if an ostensible eigenpair has a residual that is small compared to the vector of the pair, then the pair is an exact eigenpair of a small perturbation of the matrix. For example, if $\|r\|_2/\|x\|_2 = 10^{-10}$ and the elements of $A$ are all about one in magnitude, then the pair $(x, \mu)$ is an eigenpair of a matrix that differs from $A$ in the tenth place of its elements. If the elements of $A$ are accurate to, say, seven figures, it is pointless to try to compute a more accurate approximation to the eigenpair. Thus the bound can be used to terminate an iterative method for finding an eigenpair of $A$.

## The Rayleigh quotient

5. Many algorithms for the eigenproblem attempt to compute an eigenvector and do not directly produce an approximate eigenvalue. Translated into the terminology of our backward perturbation result, this means that we are given only the vector in the pair $(\mu, x)$. It is therefore natural to look for the value of $\mu$ that minimizes

$$\|r\|_2 = \|Ax - \mu x\|_2$$

and hence the norm of the backward perturbation.

6. The minimization of $\|Ax - \mu x\|_2$ is a least squares problem, with $x$ playing the role of the least squares matrix and $Ax$ playing the role of the vector to be approximated. The normal equation for $\mu$ is

$$(x^{\mathrm{H}}x)\mu = x^{\mathrm{H}}(Ax).$$

Hence

$$\mu = \frac{x^{\mathrm{H}}Ax}{x^{\mathrm{H}}x}.$$

## The power method

16. General methods for finding eigenvalues and eigenvectors are necessarily iterative. To see this, let

$$p(t) = c_0 + c_1 t + c_2 t^2 + \cdots + t^n,$$

and consider the *companion matrix*

$$C_p = \begin{pmatrix} -c_{n-1} & -c_{n-2} & \cdots & -c_1 & -c_0 \\ 1 & 0 & \cdots & 0 & 0 \\ 0 & 1 & \cdots & 0 & 0 \\ \vdots & \vdots & & \vdots & \vdots \\ 0 & 0 & \cdots & 1 & 0 \end{pmatrix}.$$

It is easy to show that the characteristic polynomial of $C_p$ is $p$. Consequently, any algorithm for solving the general eigenvalue problem can be used to solve polynomial equations.

In the nineteenth century the Norwegian mathematician Niels Abel showed that no polynomial of degree five could be solved by a finite number of additions, multiplications, divisions, and root extractions. If we had a finite algorithm for finding eigenvalues of general matrices, we could apply it to companion matrices and make a fool out of Abel. Abel was no fool.

17. Our first iterative method, the *power method*, attempts to approximate an eigenvector of $A$. Let $u_0$ be given. Having computed $u_{k-1}$, we compute $u_k$ according to the formula

$$u_k = \sigma_{k-1} A u_{k-1}. \tag{14.6}$$

Here $\sigma_{k-1}$ is a nonzero scaling factor. In practice, $\sigma_{k-1}$ will be a normalizing constant, like $\|Au_{k-1}\|_2^{-1}$, chosen to keep the components of the $u_k$ within the range of the machine. But since multiplication by a nonzero constant does not change direction of a vector, we can choose any nonzero value for $\sigma_k$. We will take advantage of this freedom in analyzing the convergence of the method.

It follows immediately from (14.6) that

$$u_k = \sigma_{k-1} \cdots \sigma_0 A^k u_0. \tag{14.7}$$

18. To analyze the convergence of the power method, suppose that we can order the eigenvalues of $A$ so that

$$|\lambda_1| > |\lambda_2| \geq |\lambda_3| \geq \cdots \geq |\lambda_n|.$$

(If this is possible, we call $\lambda_1$ the *dominant eigenvalue* of $A$.) Since $\lambda_1$ is simple, we can find a matrix

$$X = (x_1 \ X_2)$$

such that

$$X^{-1}AX = \begin{pmatrix} \lambda_1 & 0 \\ 0 & B \end{pmatrix}$$

(see §13.7). Note that $x_1$ is the eigenvector of $A$ corresponding to $\lambda_1$. Moreover, if we partition

$$X^{-1} = \begin{pmatrix} y_1^H \\ Y_2^H \end{pmatrix},$$

then $y_2^H$ is the corresponding left eigenvector.

Now let

$$v_k = X^{-1}u_k = \begin{pmatrix} \nu_k \\ w_k \end{pmatrix}.$$

Then from (14.7) we have

$$\begin{pmatrix} \nu_k \\ w_k \end{pmatrix} = \sigma_{k-1}\cdots\sigma_0 \begin{pmatrix} \lambda_1^k & 0 \\ 0 & B^k \end{pmatrix} \begin{pmatrix} \nu_0 \\ w_0 \end{pmatrix}.$$

If we choose $\sigma_i = \lambda_1$ ($i = 0,\ldots,k-1$), this equation becomes

$$\begin{pmatrix} \nu_k \\ w_k \end{pmatrix} = \begin{pmatrix} \nu_0 \\ (B/\lambda_1)^k w_0 \end{pmatrix}.$$

Multiplying by $X$, we find that

$$u_k = \nu_0 x_1 + X_2(B/\lambda_1)^k w_0. \qquad (14.8)$$

If $\nu_0 \neq 0$, the first term in the sum (14.8) is a constant multiple of the eigenvector $x_1$. The second term contains the $k$th power of $B/\lambda_1$, whose spectral radius is $|\lambda_2/\lambda_1| < 1$. Hence $(B/\lambda_1)^k \to 0$ and the iterates converge to $x_1$. We may summarize as follows.

---

Let the eigenvalues of $A$ satisfy

$$|\lambda_1| > |\lambda_2| \geq |\lambda_3| \geq \cdots \geq |\lambda_n|,$$

and let $x_1$ and $y_1^H$ be right and left eigenvectors corresponding to $\lambda_1$. If

$$y_1^H u_0 \neq 0,$$

the sequence of vectors generated by the recursion

$$u_k = \sigma_{k-1}Au_{k-1}, \qquad \sigma_k \neq 0,$$

converge (in direction) to $x_1$. The rate of convergence is asymptotically faster than the convergence of $(|\lambda_2/\lambda_1| + \epsilon)^k$ to zero for any $\epsilon > 0$.

---

The bound on the rate of convergence follows from the considerations of §14.13. For nondefective matrices we can drop the $\epsilon$ [see (14.2)].

19.     If $y_1^H u_0 \neq 0$ we say that $u_0$ *has a nonzero component along* $x_1$. This condition, which is necessary for the power method to converge to $x_1$, cannot be verified directly, since we do not know $y_1^H$. Moreover, it is not enough that $y^H u_0$ be simply nonzero — it must be substantial for the method to converge in a reasonable number of iterations. However, if $u_0$ is chosen at random, $y^H u_0$ will almost always be reasonably large.[18]

20.    The power method is recommended by its simplicity. Each iteration requires only the formation of the product of $A$ with a vector and a subsequent scaling. In forming the product, we can take advantage of zero elements in $A$ to reduce the work. As soon as we have computed $Ax_{k-1}$, we can cheaply compute a Rayleigh quotient and residual for $x_{k-1}$ and then test for convergence.

21.    But the power method has important drawbacks. It can find only a dominant eigenvalue, and it can easily happen that such an eigenvalue fails to exist. For example, if $A$ is real and has a complex eigenvalue whose magnitude is equal to its spectral radius, then the conjugate eigenvalue has the same magnitude. Even when there is a dominant eigenvalue, if it is close to the next largest eigenvalue in magnitude, the convergence will be slow — often heartbreakingly so. Although there are fixes involving iterating with more than one vector, the power method and its variants must be considered to be a special-purpose tool, not an all-purpose algorithm.

---

[18]Unless you know a lot about your problem, you should avoid systematic choices of $u_0$. For example, it very often happens that the first or last component of the eigenvector $y_1^T$ is small. In the first case the choice $\mathbf{e}_1$ will retard convergence; in the second, $\mathbf{e}_n$ will be the villain.

# Eigensystems

The Inverse Power Method
Derivation of the QR Algorithm
Local Convergence Analysis
Practical Considerations
Hessenberg Matrices

## The inverse power method

1. One cure for the difficulties of the power method is to transform the matrix so that the eigenvalues are better situated. For example, a shift of the form $A - \mu I$ may improve the ratio of the dominant eigenvalue to the subdominant eigenvalue. However, the real payoff comes when we combine a shift with an inversion.

2. Let $\mu$ be an approximation to a simple eigenvalue $\lambda_1$ of $A$. We have seen (§12.26) that the eigenvalues of $(A - \mu I)^{-1}$ are $(\lambda_i - \mu)^{-1}$. Now as $\mu$ approaches $\lambda_1$ the number $(\lambda_1 - \mu)^{-1}$ approaches infinity, while the numbers $(\lambda_i - \mu)^{-1}$ $(i \neq 1)$ will approach fixed limits. It follows that as $\mu$ approaches $\lambda_1$, the power method will converge ever more rapidly to the corresponding eigenvector.

Figure 15.1 illustrates how powerful this effect can be, even when $\mu$ is not a particularly good approximation to $\lambda$. The $*$'s in the plot represent the eigenvalues of a matrix of order five. The $+$ represents an approximation $\mu$ to the dominant eigenvalue. The o's represent shifted-and-inverted eigenvalues. Note how the dominant eigenvalue (now at five) has separated itself from the rest, which now huddle together near the origin. In particular, if we were to apply the power method to the original matrix we would have a convergence ratio of about 0.78. After shifting and inverting this ratio becomes 0.07 — more than a ten-fold speedup.

3. The power method applied to a shifted, inverted matrix is called the *inverse power method*. Of course we do not actually invert the matrix; instead we solve a linear system. Given a starting vector $u$ and an approximate eigenvalue $\mu$, one step of the inverse power method goes as follows.

$$\begin{aligned} &1. \quad \text{Solve the system } (A - \mu I)v = u \\ &2. \quad \hat{u} = v/\|v\|_2 \end{aligned} \tag{15.1}$$

The processes may be iterated by replacing $u$ by $\hat{u}$. The normalization by the 2-norm is not obligatory — any of the standard norms would do as well.

4. Since the inverse power method is a variant of the power method, the above analysis applies unchanged. However, one objection to the method must be

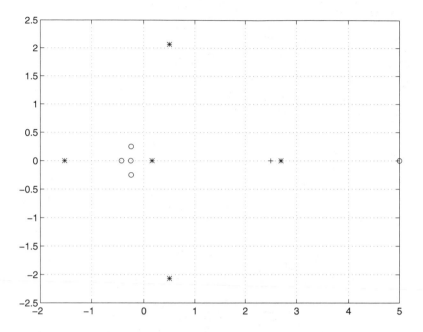

Figure 15.1. *Reciprocal, shifted eigenvalues.*

addressed in detail. If we choose $\mu$ very near to an eigenvalue, the system $(A - \mu I)v = u$ will be ill conditioned and will be solved inaccurately.

The answer to this objection is that the system is indeed solved inaccurately but that it does not matter. For if we use a stable algorithm to solve the system, then the computed solution satisfies

$$[(A + E) - \mu I]v = u,$$

where $\|A\|/\|E\|$ is of the same order of magnitude as the rounding unit. If the target eigenpair is not ill conditioned, it will not be much perturbed by the error $E$, and the method will converge.

Another way of putting it is to observe that the errors in the solution tend to lie along the direction of the eigenvector we are seeking and disappear when we normalize $v$. The failure to grasp this fact has resulted in a lot of misguided papers.

5. The size of the intermediate vector $v$ gives a lower bound on the quality of the solution. For if $\|v\|_2$ is large, the residual

$$(A - \mu I)\hat{u} = \frac{u}{\|v\|_2}$$

is small, and by the result of §14.3 the pair $(\hat{u}, \mu)$ is an exact eigenpair of a nearby matrix.

6. The advantages of the inverse power method are its swift convergence — in many applications, one iteration is enough — and the fact that it can get at eigenvalues that are surrounded by others. Its drawbacks are that it requires a good approximation to the eigenvalue whose eigenvector is sought and that one must solve the system $(A - \mu I)v = u$. For a dense matrix this solution requires $O(n^3)$ operations, although once a factorization has been computed subsequent iterations require only $O(n^2)$ operations. For special matrices, however, the expense may be considerably less.

## Derivation of the QR algorithm

7. The QR algorithm is an iterative method for computing a Schur decomposition of a matrix $A$. Starting with $A_0 = A$, it proceeds to reduce $A$ iteratively to triangular form by a sequence of unitary similarities:

$$A_{k+1} = Q_k^H A_k Q_k, \qquad k = 0, 1, 2, \ldots.$$

Although the formulas by which the $Q_k$ are generated are easy to write down, they are not in themselves very informative. We will therefore derive the formulas by exploiting the connection of the algorithm with the inverse power method. For the nonce we will be considering only one step of the iteration, and we will drop the subscript $k$.

8. The QR algorithm is an iterative version of the reduction to Schur form (see §12.31). But instead of putting zeros in the last $n-1$ elements of the first row, we try to make the first $n-1$ elements of the last row small. Specifically, write $A$ in the form

$$A = \begin{pmatrix} B & h \\ g^H & \mu \end{pmatrix}. \qquad (15.2)$$

Then the transformations $Q$ are chosen to reduce the norm of $g^H$. Once $g^H$ is effectively zero, we can continue with the process with the smaller matrix $B$, a process known as *deflation*.

9. Now an ideal choice of $Q$ would be a matrix of the form

$$Q = (Q_* \; q),$$

where $q$ is a left eigenvector of $A$ corresponding to, say, the eigenvalue $\lambda$. For in that case (just as in the reduction to Schur form),

$$Q^H A Q = \begin{pmatrix} Q_*^H A Q_* & Q_*^H A q \\ 0 & \lambda \end{pmatrix}.$$

We can then proceed immediately to the reduction of $Q_*^H A Q_*$.

10.  Lacking an exact eigenvector, we might try to compute an approximate eigenvector to use in defining $Q$. If in (15.2) the vector $g^H$ is small, the vector $\mathbf{e}_n^T$ is an approximate left eigenvector of $A$ with Rayleigh quotient $\mu$. Now it would not do to take $q = \mathbf{e}_n$ because $Q$ would then assume the form

$$Q = \begin{pmatrix} Q_* & 0 \\ 0 & 1 \end{pmatrix},$$

where $Q_*$ is unitary, and the transformation $Q^H A Q$ would replace $g^H$ by $g^H Q_*$, leaving the norm of $g^H$ unchanged. Instead, we will obtain $q^H$ by performing one step of the inverse power method with starting vector $\mathbf{e}_n^T$. The natural candidate for the shift is the Rayleigh quotient $\mu$. But because other shifts are used in some variants of the algorithm, we will shift by an unspecified quantity $\kappa$.

Since we are trying to approximate a left eigenvector, the algorithm (15.1) takes the form

1.    Solve the system $q^H(A - \kappa I) = \mathbf{e}_1^T$
2.    $\hat{q} = q/\|q\|_2$

Thus we obtain our vector $q^H$ from the formula

$$q^H = \frac{\mathbf{e}_n^T(A - \kappa I)^{-1}}{\|\mathbf{e}_n^T(A - \kappa I)^{-1}\|_2}. \tag{15.3}$$

11.  The heart of the QR algorithm is an elegant way of generating the entire matrix $Q = (Q_* \ q)$ in one fell swoop. Specifically, let

$$A - \kappa I = QR \tag{15.4}$$

be the QR decomposition of $A - \kappa I$. Then

$$\mathbf{e}_n^T R = \rho_{nn} \mathbf{e}_n^T. \tag{15.5}$$

Now write (15.4) in the form

$$Q^H = R(A - \kappa I)^{-1}.$$

Multiplying by $\mathbf{e}_n^T$, we get

$$q^H \equiv \mathbf{e}_n^T Q^H = \rho_{nn} \mathbf{e}_n^T(A - \kappa I)^{-1}.$$

Since $\|q\|_2 = 1$, it follows that the last row of the unitary matrix $Q^H$ is the vector $q^H$ in (15.3).

12.  Having generated $Q$, there is an equally elegant way to effect the similarity transformation $Q^H A Q$. From (15.4) it follows that

$$RQ = Q^H(A - \kappa I)Q = Q^H A Q - \kappa I.$$

Hence

$$Q^H AQ = RQ + \kappa I.$$

In other words, we multiply the factors $Q$ and $R$ in reverse order and add back in the shift.

13. Combining all this, we arrive at the basic formulas for the QR algorithm.[19] Given $A_0$ iterate as follows.

1.   **for** $k = 0, 1, 2, \ldots$
2.       Choose a shift $\kappa_k$
3.       Factor $A_k - \kappa_k I = Q_k R_k$, where $Q_k$ is unitary and
          $\mathbf{e}_n^T R_k = \rho_{nn}\mathbf{e}_n$                                         (15.6)
4.       $A_{k+1} = RQ + \kappa_k I$
5.   **end for** $k$

## Local convergence analysis

14. We now turn to the convergence properties of the algorithm. Global convergence can be established only in very special cases. Consequently, we will give a local convergence proof under reasonable simplifying assumptions. Once again we will drop the subscript $k$.

15. The key to the convergence analysis is to work with a partitioned form of the relation of $A - \kappa I = QR$. Specifically, let us write

$$A - \kappa I \equiv \begin{pmatrix} B - \kappa I & h \\ g^H & \mu - \kappa \end{pmatrix} = \begin{pmatrix} P & f \\ e^H & \pi \end{pmatrix} \begin{pmatrix} S & r \\ 0 & \rho \end{pmatrix} \equiv QR, \qquad (15.7)$$

and also write the relation $\hat{A} - \kappa I = RQ$ in the form

$$\hat{A} - \kappa I \equiv \begin{pmatrix} \hat{B} - \kappa I & \hat{h} \\ \hat{g}^H & \hat{\mu} - \kappa \end{pmatrix} = \begin{pmatrix} S & r \\ 0 & \rho \end{pmatrix} \begin{pmatrix} P & f \\ e^H & \pi \end{pmatrix} = RQ. \qquad (15.8)$$

We want to derive a bound for $\|\hat{g}\|_2$ in terms of $\|g\|_2$, preferably a bound that will say when the former is smaller than the latter. We proceed in stages.

16. Since $Q$ is unitary, the norms of its last row and column must be one. Hence $\|e\|_2^2 + \pi^2 = \|f\|_2^2 + \pi^2 = 1$, and

$$\|e\|_2 = \|f\|_2. \qquad (15.9)$$

---

[19] The QR *algorithm* should be carefully distinguished from the QR *decomposition*. Actually, the name of the decomposition derives from the name of algorithm — not vice versa. The "R" is from *rechts triangular* — German for "upper triangular." There is a persistent rumor that J. G. F. Francis, the inventor of the QR algorithm, originally wrote $OR$ for the factorization ($O$ for orthogonal) and later replaced the $O$ with $Q$ because the former could be confused with zero.

Next we want to show that $e$ is small when $g$ is small. Computing the $(2,1)$-block of (15.7), we get $g^H = e^H S$. If we assume that $S$ is nonsingular and set

$$\sigma = \|S^{-1}\|_2,$$

then

$$\|e\|_2 \le \sigma\|g\|_2. \tag{15.10}$$

We now need a bound on $\rho$. If we use the fact that $Q$ is unitary to rewrite (15.7) in the form

$$\begin{pmatrix} P^H & e \\ f^H & \pi \end{pmatrix} \begin{pmatrix} B - \kappa I & h \\ g^H & \mu - \kappa \end{pmatrix} = \begin{pmatrix} S & r \\ 0 & \rho \end{pmatrix},$$

then we find that $\rho = f^T h + \pi(\mu - \kappa)$. Hence from (15.9) and (15.10) and the fact that $|\pi| \le 1$,

$$\rho \le \sigma\|g\|_2\|h\|_2 + |\mu - \kappa|. \tag{15.11}$$

Finally, from (15.8) we find that $\hat{g}^H = \rho e^H$. Hence from (15.10) and (15.11) we get

$$\|\hat{g}\|_2 \le \sigma^2\|h\|_2\|g\|_2^2 + \sigma|\mu - \kappa|\|g\|_2$$

or, with the subscripts restored,

$$\|g_{k+1}\|_2 \le \sigma_k^2\|h_k\|_2\|g_k\|_2^2 + \sigma_k|\mu_k - \kappa_k|\|g_k\|_2.$$

This is our final bound.

17. The first thing this bound suggests is that we should take $\kappa_k = \mu_k$ to make the second term on the right equal to zero. Our bound then becomes

$$\|g_{k+1}\|_2 \le \eta\sigma^2\|g_k\|_2^2.$$

The shift $\kappa = \mu$ called a *Rayleigh quotient shift* because $\mu$ is the Rayleigh quotient of the vector $\mathbf{e}_n$. We will analyze the convergence of this method.

18. To help the analysis along, we will assume that

$$\sigma \text{ is a uniform bound on } \|S_k^{-1}\|_2$$

and

$$\eta \text{ is a uniform bound on } \|h_k\|_2.$$

The second assumption is always satisfied for $\eta = \|A\|_2$. The first is reasonable if we are converging to a simple eigenvalue $\lambda$. For in that case, the shifts $\kappa_k$ are approaching $\lambda$ and the $S_k$ converge to the leading principal submatrix of the R-factor of $A - \lambda I$—i.e., they do not change by much. Moreover, because $\lambda$ is simple, the inverse of the $S_k$ will be bounded.

19. Now suppose that we arrive at a point where $\eta\sigma^2\|g_k\| < 1$. Then

$$\|g_{k+1}\|_2 \le (\eta\sigma^2\|g_k\|_2)\|g_k\|_2 < \|g_k\|_2.$$

On the next iteration we have

$$\|g_{k+2}\|_2 \le (\eta\sigma^2\|g_k\|_2)^2\|g_k\|_2$$

and in general

$$\|g_{k+j}\| \le (\eta\sigma^2\|g_k\|_2)^j\|g_k\|_2.$$

Hence the $g_k$ converge to zero; i.c., the QR algorithm converges.

20. The converging iterates satisfy

$$\|g_{k+1}\| \le (\eta\sigma^2)\|g_k\|_2^2. \tag{15.12}$$

Convergence of this kind is called quadratic, and it is very fast indeed. If, for example, $\eta\sigma^2 = 1$ and $\|g_0\|_2 = 0.1$, the bounds on the subsequent iterates are $10^{-2}, 10^{-4}, 10^{-8}, 10^{-16}, \ldots$.

21. If $A$ is Hermitian, convergence occurs with blinding speed. For in that case $g = h$, and we can take $\eta = \|g_k\|$ in (15.12). Hence

$$\|g_{k+1}\| \le \sigma^2\|g_k\|_2^3.$$

If $\sigma = 1$ and $\|g_0\| = 0.1$, the bounds on the subsequent iterates are $10^{-3}, 10^{-9}, 10^{-27}, 10^{-91}, \ldots$. This kind of convergence is called cubic convergence.

## Practical considerations

22. Although the QR algorithm has excellent convergence properties, it is not as it stands a practical way to compute eigenvalues. In the first place, the algorithm requires that we compute the QR factorization of $A_k$. For a full matrix, this is an $O(n^3)$ process and would lead to an $O(n^4)$ algorithm. We need to get the count down to a more reasonable $O(n^3)$.

A second problem concerns real matrices. Since the QR algorithm is an implementation of the inverse power method with shift $\kappa$, it is imperative that $\kappa$ be near the eigenvalue we are currently seeking. But if that eigenvalue is complex, we must use a complex shift, in which case the iterates $A_k$ will become complex, and the work to implement the algorithm will more than double. Moreover, because of rounding error the eigenvalues will not occur in exact conjugate pairs.

23. A cure for the first problem is to first reduce the matrix to a special form — called Hessenberg form — that is cheap to factor and is preserved by the iteration. The second problem is solved by calculating the real Schur form

described in §13.1. The techniques by which the QR algorithm can be adapted
to compute the real Schur form are connected with the properties of Hessenberg
matrices, which we turn to now.

24. Since our final goal is to devise an algorithm for real matrices, from now
on we will assume that our hero $A$ is real.

## Hessenberg matrices

25. A matrix $A \in \mathbf{R}^{n \times n}$ is *upper Hessenberg* if

$$i > j + 1 \implies a_{ij} = 0.$$

An upper Hessenberg matrix of order 5 has a Wilkinson diagram of the form

$$
\begin{matrix}
X & X & X & X & X \\
X & X & X & X & X \\
0 & X & X & X & X \\
0 & 0 & X & X & X \\
0 & 0 & 0 & X & X
\end{matrix}
\quad . \tag{15.13}
$$

26. The first thing that we need to establish is that the QR algorithm preserves
Hessenberg form. We will prove it by describing the reduction $Q^{\mathrm{T}} A = R$ and
then the back multiplication $RQ$. To effect the reduction we will use little
$2 \times 2$ Householder transformations that act on two rows or two columns of the
matrix.[20]

Let $A$ have the form (15.13). The first step of the reduction involves
forming the product $H_1 A$ which operates on the first two rows and introduces
a zero in the $(2,1)$-element. We represent this symbolically by the diagram

$$
\begin{matrix}
\rightarrow & X & X & X & X & X \\
\rightarrow & \hat{X} & X & X & X & X \\
 & 0 & X & X & X & X \\
 & 0 & 0 & X & X & X \\
 & 0 & 0 & 0 & X & X
\end{matrix}
\quad .
$$

Here the arrows point to the rows operated on by the transformations, and
the hat indicates the element that will be annihilated. For the second step we
compute $H_2(H_1 A)$ as follows:

$$
\begin{matrix}
 & X & X & X & X & X \\
\rightarrow & 0 & X & X & X & X \\
\rightarrow & 0 & \hat{X} & X & X & X \\
 & 0 & 0 & X & X & X \\
 & 0 & 0 & 0 & X & X
\end{matrix}
\quad .
$$

---

[20] In computational practice we would work with a more convenient class of transformations
called plane rotations. To avoid interrupting the present exposition, we will introduce them
later in §18.13.

The third step computes $H_3(H_2H_1A)$:

$$
\begin{array}{c}
\\
\\
\rightarrow \\
\rightarrow \\
\\
\end{array}
\begin{array}{ccccc}
\text{X} & \text{X} & \text{X} & \text{X} & \text{X} \\
0 & \text{X} & \text{X} & \text{X} & \text{X} \\
0 & 0 & \text{X} & \text{X} & \text{X} \\
0 & 0 & \hat{\text{X}} & \text{X} & \text{X} \\
0 & 0 & 0 & \text{X} & \text{X}
\end{array}
\; .
$$

The final step computes $H_4(H_3H_2H_1A)$:

$$
\begin{array}{c}
\\
\\
\\
\rightarrow \\
\rightarrow
\end{array}
\begin{array}{ccccc}
\text{X} & \text{X} & \text{X} & \text{X} & \text{X} \\
0 & \text{X} & \text{X} & \text{X} & \text{X} \\
0 & 0 & \text{X} & \text{X} & \text{X} \\
0 & 0 & 0 & \text{X} & \text{X} \\
0 & 0 & 0 & \hat{\text{X}} & \text{X}
\end{array}
\; .
$$

The result is the triangular matrix $R = H_4H_3H_2H_1A$. The matrix $Q^{\mathrm{T}}$ is the product $H_4H_3H_2H_1$.

We must now compute $RQ = RH_1H_2H_3H_4$. The product $RH_1$ is represented symbolically thus:

$$
\begin{array}{ccccc}
\downarrow & \downarrow & & & \\
\text{X} & \text{X} & \text{X} & \text{X} & \text{X} \\
\hat{0} & \text{X} & \text{X} & \text{X} & \text{X} \\
0 & 0 & \text{X} & \text{X} & \text{X} \\
0 & 0 & 0 & \text{X} & \text{X} \\
0 & 0 & 0 & 0 & \text{X}
\end{array}
\; .
$$

Here the arrows point to the columns that are being combined, and the hat is over a zero element that becomes nonzero. The subsequent multiplications proceed as follows:

$$
(RH_1)H_2 :
\begin{array}{ccccc}
& \downarrow & \downarrow & & \\
\text{X} & \text{X} & \text{X} & \text{X} & \text{X} \\
\text{X} & \text{X} & \text{X} & \text{X} & \text{X} \\
0 & \hat{0} & \text{X} & \text{X} & \text{X} \\
0 & 0 & 0 & \text{X} & \text{X} \\
0 & 0 & 0 & 0 & \text{X}
\end{array}
\; ,
$$

$$
(RH_1H_2)H_3 :
\begin{array}{ccccc}
& & \downarrow & \downarrow & \\
\text{X} & \text{X} & \text{X} & \text{X} & \text{X} \\
\text{X} & \text{X} & \text{X} & \text{X} & \text{X} \\
0 & \text{X} & \text{X} & \text{X} & \text{X} \\
0 & 0 & \hat{0} & \text{X} & \text{X} \\
0 & 0 & 0 & 0 & \text{X}
\end{array}
\; ,
$$

$$
(RH_1H_2)H_3: \quad
\begin{array}{ccccc}
 & & & \downarrow & \downarrow \\
X & X & X & X & X \\
X & X & X & X & X \\
0 & X & X & X & X \\
0 & 0 & X & X & X \\
0 & 0 & 0 & \hat{0} & X
\end{array}
\; .
$$

The final matrix is in Hessenberg form as required.

27. The algorithm is quite efficient. The product $H_1A$ is effectively the product of a $2 \times 2$ matrix with the first two rows of $A$. Since these rows are of length $n$ the process requires about $2n$ additions and $4n$ multiplications. The second product combines two rows of length $n-1$ and requires about $2(n-1)$ additions and $4(n-1)$ multiplications. Thus the number of additions for the reduction to triangular form is

$$
2\sum_{i=1}^{n} i \cong n^2.
$$

Similarly the number of multiplications is $2n^2$. The back multiplication has the same operation count. Consequently the total count is $2n^2$ additions and $4n^2$ multiplications.

# Eigensystems

Reduction to Hessenberg form
The Hessenberg QR Algorithm
Return to Upper Hessenberg

## Reduction to Hessenberg form

1. We have seen that the QR algorithm preserves Hessenberg form and can be implemented in $O(n^2)$ operations. But how do we get to a Hessenberg matrix in the first place? We will now describe a reduction to Hessenberg form by Householder transformations. The reduction bears a passing similarity to the reduction to triangular form in §8.14; however, the fact that we must work with similarity transformations prevents us from getting all the way to triangularity.

2. We will begin in the middle of things. Suppose we have reduced our matrix $A$ to the form illustrated for $n = 8$ in the following Wilkinson diagram:

$$\begin{pmatrix} X & X & X & X & X & X & X & X \\ X & X & X & X & X & X & X & X \\ 0 & X & X & X & X & X & X & X \\ 0 & 0 & X & X & X & X & X & X \\ \hline 0 & 0 & 0 & X & X & X & X & X \\ 0 & 0 & 0 & \widehat{X} & X & X & X & X \\ 0 & 0 & 0 & \widehat{X} & X & X & X & X \\ 0 & 0 & 0 & \widehat{X} & X & X & X & X \end{pmatrix}. \tag{16.1}$$

We generate a Householder transformation of the form $\operatorname{diag}(I_4, H)$ to eliminate the elements with hats on them. The resulting matrix $\operatorname{diag}(I_4, H)A\operatorname{diag}(I_4, H)$ is then one step nearer to Hessenberg form.

3. Northwest indexing will help us keep our subscripts straight (see §6.1). Let us partition the current matrix $A$ in the form

$$\begin{pmatrix} A_{11} & a_{1k} & A_{1,k+1} \\ 0 & a_{k+1,k} & A_{k+1,k+1} \end{pmatrix}.$$

For $k = 4$, this partition corresponds to the partitioning of (16.1). We now generate a Householder transformation of the form $\operatorname{diag}(I_k, H_k)$, where $H_k a_{k+1,k} = \alpha e_1$. It follows that

$$\begin{pmatrix} I_k & 0 \\ 0 & H_k \end{pmatrix} \begin{pmatrix} A_{11} & a_{1k} & A_{1,k+1} \\ 0 & a_{k+1,k} & A_{k+1,k+1} \end{pmatrix} \begin{pmatrix} I_k & 0 \\ 0 & H_k \end{pmatrix}$$

$$= \begin{pmatrix} A_{11} & a_{1k} & A_{1,k+1}H_k \\ 0 & \alpha e_1 & H_k A_{k+1,k+1} H_k \end{pmatrix}.$$

The process continues with the matrix $H_k A_{k+1,k+1} H_k$.

4. The body of the loop in the following algorithm consists of a Householder generation, a premultiplication by the resulting $H$, and a postmultiplication by $H$. It also accumulates the transformations (see §16.10).

$$
\begin{array}{ll}
1. & U = I \\
2. & \textbf{for } k = 1 \textbf{ to } n-2 \\
3. & \quad housegen(A[k+1{:}n, k], u, A[k+1, k]) \\
4. & \quad v^{\mathrm{T}} = u^{\mathrm{T}} * A[k+1{:}n, k+1{:}n] \\
5. & \quad A[k+1{:}n, k+1{:}n] = A[k+1{:}n, k+1{:}n] - u * v^{\mathrm{T}} \\
6. & \quad v = A[1{:}n, k+1{:}n] * u \\
7. & \quad A[1{:}n, k+1{:}n] = A[1{:}n, k+1{:}n] - v * u^{\mathrm{T}} \\
8. & \quad v = U[1{:}n, k+1{:}n] * u \\
9. & \quad U[1{:}n, k+1{:}n] = U[1{:}n, k+1{:}n] - v * u^{\mathrm{T}} \\
10. & \textbf{end for } i
\end{array}
\tag{16.2}
$$

Note that in the postmultiplication (statements 6 and 7), the transformation is applied to every row of the matrix from 1 to $n$.

5. If we exclude the accumulation of the Householder transformations, the algorithm requires $\frac{5}{3}n^3$ additions and multiplications. On the other hand, one iteration of the QR algorithm applied to a full matrix costs $\frac{4}{3}n^3$ additions and multiplications. Thus for what is essentially the cost of a single iteration you reduce the cost of all subsequent iterations to $O(n^2)$—a good bargain.

6. The algorithm is stable in the sense that the final Hessenberg form is the one that would be computed by exact computations on a matrix $A + E$, where $\|E\|_2/\|A\|_2$ is of the order of the rounding unit of the machine in question. It should not be thought that this stability result means that the final Hessenberg form—call it $B$—is accurate. In fact, most of $B$ could be totally different from the matrix you would get by exact calculation. But because it is exactly similar to $A + E$, it has accurate copies of the well-conditioned eigenvalues of $A$.

7. It is natural to ask why the algorithm cannot be modified to triangularize $A$ by eliminating the elements $A[k{:}n, k]$ instead of the elements $A[k+1{:}n, k]$. The reason is that the transformation then combines rows $k$ through $n$, and the back multiplication would combine columns $k$ through $n$, destroying the zeros just introduced in column $k$. Thus the transformation must be offset from the diagonal, and that offset causes the product of all the transformations to assume the form

$$
U = \begin{pmatrix} 1 & 0 \\ 0 & U_* \end{pmatrix}.
$$

We will use this result in §17.2 to derive the QR algorithm with implicit double shift.

## The Hessenberg QR algorithm

8. If we are willing to put up with complex arithmetic, we now have a satisfactory algorithm for computing a Schur form of a general matrix. First reduce the matrix to Hessenberg form. Then apply the QR algorithm with Rayleigh quotient shifts — except for the first which should be complex to move the iteration into the complex plane. Typically, the algorithm will flounder for a few iterations and then start converging. When $a_{n,n-1}$ becomes sufficiently small (say compared with $|a_{nn}| + |a_{n-1,n-1}|$), we can set it to zero and continue by using $a_{n-1,n-1}$ as a shift. It turns out that as the subdiagonal element next to the current shift is converging quadratically to zero, the other subdiagonal elements also tend slowly to zero.[21] Consequently, as we move up the matrix, it takes fewer and fewer iterations to find an eigenvalue.

9. There is a subtlety about applying the transformations that is easy to overlook. Suppose that we have reduced the matrix to the form

$$
\begin{pmatrix}
a & a & a & a & a & a & a & a \\
a & a & a & a & a & a & a & a \\
0 & a & a & a & a & a & a & a \\
0 & 0 & a & a & a & a & a & a \\
0 & 0 & 0 & a & a & a & a & a \\
\hline
0 & 0 & 0 & 0 & 0 & a & a & a \\
0 & 0 & 0 & 0 & 0 & 0 & a & a \\
0 & 0 & 0 & 0 & 0 & 0 & 0 & a
\end{pmatrix},
$$

so that we have found three eigenvalues and have five to go. Then the next QR step generates four transformations. How they are applied to the matrix depends on whether we want to compute the Schur form or just find the eigenvalues of $A$. If we want to compute a Schur form, then we must apply the transformations across the entire matrix — i.e., to all the elements above the line. But if we are interested only in the eigenvalues then we need only work with the leading principal submatrix indicated by the lines below:

$$
\begin{pmatrix}
a & a & a & a & a & a & a & a \\
a & a & a & a & a & a & a & a \\
0 & a & a & a & a & a & a & a \\
0 & 0 & a & a & a & a & a & a \\
0 & 0 & 0 & a & a & a & a & a \\
\hline
0 & 0 & 0 & 0 & 0 & a & a & a \\
0 & 0 & 0 & 0 & 0 & 0 & a & a \\
0 & 0 & 0 & 0 & 0 & 0 & 0 & a
\end{pmatrix}.
$$

---

[21]This is a consequence of a relation between the unshifted QR algorithm and the power method.

Thus if we require only eigenvalues we can save work in applying the transformation.

10.  If we want the Schur form, we must accumulate the transformations that were generated during the reduction to Hessenberg form and in the course of the QR iterations. Specifically, beginning with $U = I$, each time we perform a similarity transformation of the form $A \leftarrow H^H A H$, we also perform the update $U \leftarrow UH$ [see statements 8 and 9 in (16.2)].

11.  We can compute eigenvectors from the Schur form. To compute an eigenvector corresponding to an eigenvalue $\lambda$, we partition the Schur form to expose the eigenvalue:

$$\begin{pmatrix} B & c & D \\ 0 & \lambda & e^T \\ 0 & 0 & F \end{pmatrix}.$$

We then seek an eigenvector from the equation

$$\begin{pmatrix} B & c & D \\ 0 & \lambda & ^T \\ 0 & 0 & F \end{pmatrix} \begin{pmatrix} w \\ 1 \\ 0 \end{pmatrix} = \lambda \begin{pmatrix} w \\ 1 \\ 0 \end{pmatrix}.$$

It follows that

$$(B - \lambda I)w = -c. \tag{16.3}$$

This is an upper triangular system that can be solved for $w$. The eigenvector of the original matrix is then given by

$$x = U \begin{pmatrix} w \\ 1 \\ 0 \end{pmatrix}.$$

This procedure might appear risky when $A$ has multiple eigenvalues, since $\lambda$ could then appear on the diagonal of $B$. However, rounding error will almost certainly separate the eigenvalues a little (and if it fails to, you can perturb the diagonals of $B$), so that $B - \lambda I$ is not singular. Nor is there any danger in dividing by a small element as long as the eigenvector is well conditioned. For in that case, the numerator will also be small. However, care must be taken to scale to avoid overflows in solving (16.3), just in case there is a nearly defective eigenvalue.

12.  When $A$ is Hermitian, the Hessenberg form becomes tridiagonal. Both the reduction and the QR algorithm simplify; and, as we have seen, the convergence is cubic. The details can be found in many places, but you will find it instructive to work them out for yourself.

## Return to Upper Hessenberg

13. The algorithm just described is a viable method for computing eigenvalues and eigenvectors of a general complex matrix. But with real matrices we can do better by computing the real Schur form of $A$ (§13.1) in real arithmetic. In order to derive the algorithm, we must establish a result about Hessenberg matrices that is of interest in its own right.

14. First some nomenclature. We will say that an upper Hessenberg matrix of order $n$ is *unreduced* if $a_{i+1,i} \neq 0$ ($i = 1, \ldots, n-1$). In other words, the subdiagonal elements of an unreduced upper Hessenberg matrix are nonzero.

In solving eigenvalue problems we deal principally with unreduced Hessenberg matrices. For if a matrix is reduced, say it has the form

$$\begin{pmatrix} X & X & X & X & X & X \\ X & X & X & X & X & X \\ 0 & 0 & X & X & X & X \\ 0 & 0 & X & X & X & X \\ 0 & 0 & 0 & X & X & X \\ 0 & 0 & 0 & 0 & X & X \end{pmatrix},$$

then the problem uncouples — in the illustration above into a $2 \times 2$ problem corresponding to the leading principal submatrix and a $4 \times 4$ problem corresponding to the trailing principal submatrix. It would be inefficient not to take advantage of this uncoupling.

15. The basic fact about reduction to unreduced Hessenberg form is the following:

> Let $A \in \mathbf{R}^n$. If
> $$Q^{\mathrm{T}} A Q = H,$$
> where $Q$ is unitary and $H$ is unreduced upper Hessenberg, then $Q$ and $H$ are uniquely determined by the first column of $Q$.

The proof is an algorithm, called *Arnoldi reduction*, that is reminiscent of the method for solving Sylvester's equation in §13.6. Write the equation $AQ = QH$ in the form

$$A(q_1 \ q_2 \ q_3 \ q_4 \ \cdots) = (q_1 \ q_2 \ q_3 \ q_4 \cdots) \begin{pmatrix} h_{11} & h_{12} & h_{13} & \cdots \\ h_{21} & h_{22} & h_{23} & \cdots \\ 0 & h_{32} & h_{33} & \cdots \\ 0 & 0 & h_{43} & \cdots \\ \vdots & \vdots & \vdots & \end{pmatrix}. \tag{16.4}$$

Now $h_{11} = q_1^{\mathrm{T}} A q_1$ is determined by $q_1$. Computing the first column of (16.4), we find that

$$A q_1 = h_{11} q_1 + h_{21} q_2$$

or

$$q_2 = \frac{Aq_1 - h_{11}q_1}{h_{21}},$$

where $h_{21}$ is determined by the requirement that $\|q\|_2 = 1$. It must be nonzero because $H$ is unreduced. We now have $h_{12} = q_1^\mathrm{T} Aq_2$ and $h_{22} = q_2^\mathrm{T} Aq_2$. Hence computing the second column of (16.4) we get

$$Aq_2 = h_{12}q_1 + h_{22}q_2 + h_{32}q_3,$$

from which it follows that

$$q_3 = \frac{Aq_2 - h_{12}q_1 - h_{22}q_2}{h_{32}},$$

where $h_{32}$ is a nonzero normalization constant. In general

$$h_{i,j-1}q_i^\mathrm{T} Aq_{j-1}, \qquad i = 1, \ldots, j-1, \tag{16.5}$$

and

$$q_j = \frac{Aq_{j-1} - \sum_{i=1}^{j-1} h_{i,j-1}q_i}{h_{j,j-1}}. \tag{16.6}$$

Thus a knowledge of $q_1$ enables us to compute $Q$ and $H$.

16.   Arnoldi reduction [(16.5) and (16.6)] has the flavor of Gram–Schmidt orthogonalization, and as with the latter algorithm cancellation can cause loss of orthogonality. Since we are orthogonalizing $Aq_{j-1}$ against the previous $q_i$, the algorithm also has the flavor of the recursion for generating orthogonal polynomials, although we do not get a three-term recurrence. However, if $A$ is Hermitian, so is $Q^\mathrm{H} AQ$, which is therefore tridiagonal. Thus the recursion (16.6) simplifies to

$$q_j = \frac{Aq_{j-1} - h_{j-1,j-1}q_{j-1} - h_{j-2,j-1}q_{j-2}}{h_{j,j-1}},$$

which is an actual three-term recurrence. We will return to these points when we discuss the Arnoldi and Lanczos algorithm (Lectures 19 and 20).

# Eigensystems

The Implicit Double Shift
Some Implementation Details
The Singular Value Decomposition

## The implicit double shift

1. We now turn to an ingenious technique for computing the real Schur form of a real Hessenberg matrix without performing complex arithmetic when $A$ is real. The idea is to perform two complex conjugate shifts at once. To set the stage, consider two steps of the QR algorithm:

$$
\begin{array}{llll}
1. & A_0 - \kappa I = Q_0 R_0, & 2. & A_1 = Q_0^H A_0 Q_0, \\
3. & A_1 - \bar{\kappa} I = Q_1 R_1, & 4. & A_2 = Q_1^H A_1 Q_1.
\end{array} \tag{17.1}
$$

I claim that the product $Q = Q_0 Q_1$ and hence $A_2 = Q^T A_0 Q$ is real. To see this, first note that the matrix

$$
(A_0 - \bar{\kappa} I)(A_0 - \kappa I) = A_0^2 - 2\Re(\kappa) A_0 + |\kappa|^2 I
$$

is real. But

$$
\begin{aligned}
(A_0 - \bar{\kappa} I)(A_0 - \kappa I) &= (A_0 - \bar{\kappa} I) Q_0 R_0 & \text{by (17.1.1)} \\
&= Q_0 Q_0^H (A_0 - \bar{\kappa} I) Q_0 R_0 & \text{since } QQ^H = I \\
&= Q_0 (A_1 - \bar{\kappa} I) Q_0 R_0 & \text{by (17.1.2)} \\
&= (Q_0 Q_1)(R_1 R_0) & \text{by (17.1.3)}.
\end{aligned}
$$

Hence $Q = Q_0 Q_1$ is just the Q factor of the QR factorization of $(A_0 - \bar{\kappa} I)(A_0 - \kappa I)$ and hence is real.[22]

2. Thus if we perform two steps of the QR algorithm, with complex conjugate shifts, the result is nominally real. In practice, however, rounding error will give us a complex matrix — one that can have substantial imaginary components. Fortunately, there is an indirect way to compute $Q$ and $A_2$ without going through $A_1$.

Consider the following algorithm.

$$
\begin{array}{ll}
1. & \text{Compute the first column } q \text{ of } Q \\
2. & \text{Let } H \text{ be an orthogonal matrix such that } H e_1 = q \\
3. & \text{Let } B = H^T A H \\
4. & \text{Use algorithm (16.2) to find } U \text{ such that } C = U^T B U \\
& \quad \text{is Hessenberg}
\end{array} \tag{17.2}
$$

---

[22] Well, almost. Its columns could be scaled by complex factors of magnitude one.

Now because $U$ has the form (§16.7)

$$U = \begin{pmatrix} 1 & 0 \\ 0 & U_* \end{pmatrix},$$

$HUe_1 = He_1 = q$—i.e., the matrix $HU$ is an orthogonal matrix whose first column is $q$. It also reduces $A_0$ to Hessenberg form. But $Q$ itself is an orthogonal matrix whose first column is $q$ and which reduces $A_0$ to Hessenberg form. Hence by the result of §16.15, $Q = HU$ and $C = A_2$. Thus the above algorithm computes the results of the double shift directly.

3. To get a working algorithm, we must first show how to compute the first column of $Q$ and then how to reduce $H^T A H$. Both problems are simplified by the fact that $A_0$ is Hessenberg. For notational simplicity, we will drop the subscript zero.

4. Since $Q$ is the Q-factor of a QR factorization of $(A - \bar{\kappa}I)(A - \kappa I)$, the first column of $Q$ is proportional to the first column of $(A - \bar{\kappa}I)(A - \kappa I)$. But this column is easily seen to be

$$\begin{pmatrix} a_{11} - \bar{\kappa} & a_{12} \\ a_{21} & a_{22} - \bar{\kappa} \\ 0 & a_{32} \\ 0 & 0 \end{pmatrix} \begin{pmatrix} a_{11} - \kappa \\ a_{21} \end{pmatrix}$$

$$= \begin{pmatrix} a_{11}^2 - 2\Re(\kappa)a_{11} + |\kappa|^2 + a_{12}a_{21} \\ a_{21}a_{11} + a_{22}a_{21} - 2a_{21}\Re(\kappa) \\ a_{32}a_{21} \\ 0 \end{pmatrix} \equiv \begin{pmatrix} w \\ 0 \end{pmatrix}.$$

$$(17.3)$$

5. The first column of $Q$ has only three nonzero elements. Hence if we take $H$ in (17.2) to be a Householder transformation, it will affect only the first three rows and columns of $A$. Specifically $H^T A H$ will have the form

$$\begin{pmatrix}
a & a & a & a & a & a & a & a \\
a & a & a & a & a & a & a & a \\
\hat{a} & a & a & a & a & a & a & a \\
\hat{a} & a & a & a & a & a & a & a \\
0 & 0 & 0 & a & a & a & a & a \\
0 & 0 & 0 & 0 & a & a & a & a \\
0 & 0 & 0 & 0 & 0 & a & a & a \\
0 & 0 & 0 & 0 & 0 & 0 & a & a
\end{pmatrix}.$$

The next step is to reduce the first column of this matrix, i.e., to annihilate the elements with hats. This can be done by a Householder transformation that affects only rows and columns two through four of the matrix. The transformed matrix has the form

$$
\begin{pmatrix}
a & a & a & a & a & a & a & a \\
a & a & a & a & a & a & a & a \\
0 & a & a & a & a & a & a & a \\
0 & \hat{a} & a & a & a & a & a & a \\
0 & \hat{a} & a & a & a & a & a & a \\
0 & 0 & 0 & 0 & a & a & a & a \\
0 & 0 & 0 & 0 & 0 & a & a & a \\
0 & 0 & 0 & 0 & 0 & 0 & a & a
\end{pmatrix}. \tag{17.4}
$$

We continue by using $3\times3$ Householder transformations to eliminate the hatted elements in (17.4). In this way we chase the bulge below the subdiagonal out the bottom of the matrix. At the $k$th stage, the elements $a_{k+2,k}$ and $a_{k+3,k}$ are annihilated, but two new elements are introduced into the $(k+4, k+1)$- and $(k+4, k+2)$-positions. The very last step is special, since only one element is annihilated in the matrix

$$
\begin{pmatrix}
a & a & a & a & a & a & a & a \\
a & a & a & a & a & a & a & a \\
0 & a & a & a & a & a & a & a \\
0 & 0 & a & a & a & a & a & a \\
0 & 0 & 0 & a & a & a & a & a \\
0 & 0 & 0 & 0 & a & a & a & a \\
0 & 0 & 0 & 0 & 0 & a & a & a \\
0 & 0 & 0 & 0 & 0 & \hat{a} & a & a
\end{pmatrix}.
$$

6. Figure 17.1 exhibits a program implementing this reduction. The main trick in the code is the incorporation of the initial transformation based on $u$ as a zeroth transformation in the loop. The integers $l$ and $m$ make sure the transformations stay within the array $A$. In particular, when $k = n - 2$, the transformation becomes $2 \times 2$ rather than $3 \times 3$.

## Some implementation details

7. The conventional choice of shifts are the eigenvalues of the matrix

$$
\begin{pmatrix}
a_{n-1,n-1} & a_{n-1,n} \\
a_{n,n-1} & a_{nn}
\end{pmatrix}. \tag{17.5}
$$

We do not have to calculate them, since all we need are the quantities

$$
2\Re(\kappa) = a_{n-1,n-1} + a_{nn}
$$

1.    Calculate $w$ from (17.3)
2.    **for** $k = 0$ **to** $n-2$
3.        $l = \min\{k+3, n\}$
4.        **if** $(k = 0)$
5.            $housegen(w, u, t)$
6.        **else**
7.            $housegen(A(k+1:l, k), u, A(k+1, k))$
8.            $A[k+2:l, k] = 0$
9.        **end if**
10.       $v^{\mathrm{T}} = u^{\mathrm{T}} * A[k+1:l, k+1:n]$
11.       $A[k+1:l, k+1:n] = A[k+1:l, k+1:n] - u * v^{\mathrm{T}}$
12.       $m = \min\{k+4, n\}$
13.       $v = A[1:m, k+1:l] * u$
14.       $A[1:m, k+1:l] = A[1:m, k+1:l] - v * u^{\mathrm{T}}$
15.   **end for** $k$

Figure 17.1. *Doubly shifted QR step.*

and

$$|\kappa|^2 = a_{n-1,n-1} a_{nn} - a_{n,n-1} a_{n-1,n},$$

i.e., the trace and the determinant of (17.5). These formulas also work when the eigenvalues of (17.5) are real.

If at convergence the eigenvalues of (17.5) are complex, they will form a complex conjugate pair. In this case, the element $a_{n-2,n-1}$ will converge quadratically to zero. If they are real, both $a_{n-1,n}$ and $a_{n-2,n-1}$ will converge to zero.

8. As the iteration progresses, either $a_{n,n-1}$ or $a_{n-1,n-2}$ will become small enough to be regarded as zero, and we can deflate the problem as described in §16.8. If $a_{n-1,n-2}$ has converged, one usually triangularizes the $2 \times 2$ block whenever it has real eigenvalues. This seemingly simple computation is surprisingly difficult to get right.

9. With large problems, we may find that other subdiagonal elements become

zero, so that the matrix assumes a form like the following.

$$
\begin{array}{cccccccccc}
a & a & a & a & a & a & a & a & a & a \\
a & a & a & a & a & a & a & a & a & a \\
  & a & a & a & a & a & a & a & a & a \\
  &   & 0 & a & a & a & a & a & a & a & \leftarrow i_1 \\
  &   &   & a & a & a & a & a & a & a \\
  &   &   &   & a & a & a & a & a & a \\
  &   &   &   &   & a & a & a & a & a & \leftarrow i_2 \\
  &   &   &   &   &   & 0 & a & a & a \\
  &   &   &   &   &   &   & a & a & a \\
  &   &   &   &   &   &   &   & 0 & a
\end{array}
$$

In this case it is important to confine the iteration between rows $i_1$ and $i_2$. There are two reasons. First, it saves work. Second, if one starts at the top of the matrix, the reduction to Hessenberg form stops at the first subdiagonal zero it encounters, leaving the rest of the matrix unchanged. Needless to say, this does not speed convergence at $i_2$.

As we pointed out in §16.9, if only eigenvalues are required, we need only apply the transformations to $A[i_1{:}i_2, i_1{:}i_2]$. However, if we want to compute the entire Schur form, we must premultiply the transformations into $A[i_1{:}i_2, i_1{:}n]$ and postmultiply them into $A[1{:}i_2, i_1{:}i_2]$.

After each iteration we must search back through the subdiagonal elements, starting at $i_2$, to find the first negligible one. This search can be combined with a check for convergence. It is not easy to get right.

## The singular value decomposition

10. The QR algorithm is notable for being applicable to more than garden variety eigenvalue problems. As an example, we are going to show how to use it to calculate one of the more important decompositions in matrix decompositions: the *singular value decomposition*.

11. Let $X \in \mathbf{R}^{n \times n}$. We have seen (§7.5) that there is a an orthogonal matrix $Q$ such that $Q^{\mathrm{T}}X$ is upper triangular. Now the columns of $Q$ have roughly $\frac{1}{2}k^2$ degrees of freedom — enough to reduce $X$ to triangular form but no further. If, however, we postmultiply $X$ by an orthogonal matrix, there are enough additional degrees of freedom to diagonalize $X$, even when $X$ is rectangular. The result is called the singular value decomposition.

Let $X \in \mathbf{R}^{n \times k}$ with $n \geq k$. Then there are orthogonal matrices $U$ and $V$ such that

$$U^{\mathrm{T}} X V = \begin{pmatrix} \Sigma \\ 0 \end{pmatrix}, \qquad (17.6)$$

where

$$\Sigma = \mathrm{diag}(\sigma_1, \ldots, \sigma_k) \qquad (17.7)$$

with

$$\sigma_1 \geq \sigma_2 \geq \cdots \geq \sigma_k \geq 0. \qquad (17.8)$$

The scalars $\sigma_i$ are called the *singular values* of $X$. The columns of $U$ and $V$ are called *left and right singular vectors*.

12. The decomposition was discovered independently by Beltrami in 1873 and Jordan (of canonical form fame) in 1874. Though Jordan's derivation is by far the more elegant, we will give Beltrami's, for which we have already set the stage.

13. The matrix $X^{\mathrm{T}} X$ is symmetric. Any eigenvalue of $X^{\mathrm{T}} X$ is nonnegative. For if $(X^{\mathrm{T}} X) v = \lambda v$ with $\|v\|_2 = 1$, then $\lambda = v^{\mathrm{T}}(X^{\mathrm{T}} X) v = \|X v\|_2^2 \geq 0$. Hence (§13.15) there is an orthogonal matrix $V$ such that

$$V^{\mathrm{T}}(X^{\mathrm{T}} X) V = \Sigma^2,$$

where $\Sigma$ satisfies (17.7) and (17.8).

Partition $\Sigma = \mathrm{diag}(\Sigma_1, 0)$, where the diagonal elements of $\Sigma_1$ are positive and partition $V = (V_1 \ V_2)$ conformally. Then, $V_2^{\mathrm{T}}(X^{\mathrm{T}} X) V_2 = 0$. But the diagonals of the matrix are the square norms of the columns of $X V_2$. Hence

$$X V_2 = 0. \qquad (17.9)$$

Let

$$U_1 = X V_1 \Sigma_1^{-1}. \qquad (17.10)$$

Then

$$U_1^{\mathrm{T}} U_1 = \Sigma_1^{-1} V_1^{\mathrm{T}} X^{\mathrm{T}} X V_1 \Sigma_1^{-1} = \Sigma_1^{-1} \Sigma_1^2 \Sigma_1^{-1} = I,$$

so that the columns of $U_1$ are orthonormal. Let $U = (U_1 \ U_2)$ be orthogonal. Then from (17.10)

$$U_1^{\mathrm{T}} X V_1 = U_1^{\mathrm{T}} U_1 \Sigma_1 = \Sigma_1$$

and

$$U_2^{\mathrm{T}} X V_1 = U_2^{\mathrm{T}} U_1 \Sigma_1 = 0.$$

These equations, along with (17.9), imply that $U^{\mathrm{T}} X V$ satisfies (17.6).

# Eigensystems

Rank and Schmidt's Theorem
Computational Considerations
Reduction to Bidiagonal Form
Plane Rotations
The Implicit QR Algorithm for Singular Values

## Rank and Schmidt's theorem

1. The singular value decomposition has many uses, but perhaps the most important lies in its ability to reveal the rank of a matrix. Suppose that $X$ has rank $p$. Since $U$ and $V$ are nonsingular, $\Sigma$ must also have rank $p$, or equivalently

$$\sigma_1 \geq \cdots \geq \sigma_p > \sigma_{p+1} = \cdots = \sigma_k = 0.$$

This shows that in principle we can determine the rank of a matrix by inspecting its singular values.

2. But life is not that simple. In practice, the matrix $X$ we actually observe will be a perturbation of the original matrix $X_{\text{true}}$ — say $X = X_{\text{true}} + E$. The zero singular values of $X_{\text{true}}$ will become nonzero in $X$, though it can be shown that they will be bounded by $\|E\|_2$. Thus, if the nonzero singular values of $X_{\text{true}}$ are well above $\|E\|_2$, and we compute the singular value of $X$, we will observe $p$ singular values above the error level and $n - p$ singular values at the error level. Thus we can guess the original rank of $X$.[23]

3. Having found an ostensible rank $p$ of the original matrix $X_{\text{true}}$, we would like to find an approximation of that matrix of exactly rank $p$. Since our knowledge of $X_{\text{true}}$ is derived from $X$, we would not want our approximation to stray too far from $X$. The following construction does the trick.

4. Partitioning $U$ and $V$ by columns, let

$$U_p = (u_1 \ u_2 \ \cdots \ u_p) \quad \text{and} \quad V_p = (v_1 \ v_2 \ \cdots \ v_p)$$

and let

$$\Sigma_p = \text{diag}(\sigma_1, \sigma_2, \ldots, \sigma_p).$$

Then the matrix

$$X_p = U_p \Sigma_p V_p^{\text{T}}$$

---

[23] This is an oversimplification of a hard problem. Among other things, we need to know the size of the error. It is not clear what to do about singular values that are near but above the error level. And one must decide how to scale the columns of $X$, since such scaling changes the singular values. Still the problems are not insurmountable — at least in individual cases where we can apply a dose of common sense.

is of rank not greater than $p$. And by a theorem of Schmidt,[24] it is a best approximation in the following sense:

$$\|X_p - X\|_{\mathrm{F}} \le \min_{\mathrm{rank}(Y) \le p} \|Y - X\|_{\mathrm{F}}.$$

Since the square of the Frobenius norm of $Y - X$ is the sum of squares of the differences of the elements of $Y - X$, the problem of minimizing $\|Y - X\|_{\mathrm{F}}^2$ is a nonlinear least squares problem (the set of matrices of rank $p$ or less is not a linear space). It is one of the few such problems with an explicit solution.

## Computational considerations

5.   The derivation of the singular value decomposition in §17.13 suggests an algorithm for computing the singular value decomposition.

1.   Form $A = X^{\mathrm{T}}X$
2.   Compute an eigendecomposition $V^{\mathrm{T}}AV = \Sigma^2$
3.   Partition $\Sigma = \mathrm{diag}(\Sigma_1, 0)$, where $\Sigma_1$ is nonsingular
4.   Partition $V = (V_1 \ V_2)$ conformally
5.   Set $U_1 = XV_1\Sigma_1^{-1}$
6.   Determine $U_2$ so that $U = (U_1 \ U_2)$ is orthogonal

(18.1)

Unfortunately, this algorithm has problems.

6.   In the first place, we have already noted that it is difficult to distinguish zero singular values from nonzero singular values in the presence of error. The same applies to eigenvalues. Thus the partitioning of $\Sigma$ in the third step of the algorithm forces us to make a decision that we are not really prepared to make.

7.   Second, suppose that $X$ has been scaled so that its columns have norm one and suppose that $\sigma_i$ is a small singular value of $X$. By statement 5, we must compute the corresponding left singular vector in the form

$$u_i = \sigma_i^{-1}Xv_i,$$

where $v_i$ is from the eigensystem of $A$. Since $\|u_i\|_2 = 1$, it follows that $\|Xv_i\|_2 = \sigma$; i.e., $Xv_i$ is small. Since $\|v\|_2 = 1$ and the columns of $X$ have norm one, the vector $Xv_i$ has to be computed with cancellation and will be inaccurate. These inaccuracies will propagate to $u_i$.

8.   A third problem is the squaring of the singular values as we pass from $X$ to $A$. Figure 18.1 depicts representative singular values of $X$ by white arrows

---

[24]Schmidt (of Gram–Schmidt fame) proved his result for integral equations in a groundbreaking 1907 paper that anticipated later developments in functional analysis. The result was rediscovered for matrices in 1936 by Ekart and Young and whose names are often associated with it.

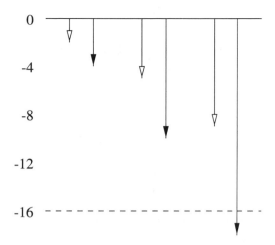

Figure 18.1. *Effect of squaring singular values.*

descending from a nominal value of 1. They are hanging over a sea of rounding error, represented by the dashed line at $10^{-16}$. The accuracy we can expect of a singular value is the distance from the arrow to the sea. The black arrows represent the eigenvalues of $A = X^{\mathrm{T}}X$. The first singular value has roughly 14 accurate digits while its square has 12. The second has 9 digits while its square has 6. The third singular value has 7 digits; its square, on the other hand, has become shark bait — no accuracy at all. Thus in passing from $A$ to $A^{\mathrm{T}}A$ we diminish the accuracy of the singular values we are trying to compute.

9.  It is important to keep some perspective on these defects. The method of computing $U$ from $V$ is certainly bad — $V$ is orthogonal to working accuracy, while $U$ is not. However, in some applications, all that is required is $V$ and $\Sigma$. If $X$ is sparse, the most efficient algorithm will be to compute $A$ and its eigensystem. The above considerations suggest that if the singular values are well above the square root of the rounding unit (assuming $\|X\|_2 \cong 1$), then the results may be satisfactory. (Compare the comments on the normal equations in §9.33.)

## Reduction to bidiagonal form

10.  As an alternative to the algorithm (18.1) we are going to show how the QR algorithm can be used to calculate singular values without forming the cross-product matrix $A$. Now an efficient implementation of the QR algorithm requires that $A$ be reduced to Hessenberg form. Since $A$ is symmetric, the Hessenberg form becomes tridiagonal — that is, a form in which all elements of $A$ above the first superdiagonal and below the first subdiagonal are zero. The corresponding form for $X$ is a *bidiagonal form,* in which only the diagonal and

the first superdiagonal are nonzero. We will now show how to compute this
form.

11. The process begins like the orthogonal triangularization algorithm of §8.14.
A Householder transformation introduces zeros into the first column as illustrated below:

$$P_1 X = \begin{pmatrix} X & X & X & X \\ 0 & X & X & X \\ 0 & X & X & X \\ 0 & X & X & X \\ 0 & X & X & X \\ 0 & X & X & X \end{pmatrix}.$$

Next the matrix is postmultiplied by a Householder transformation that introduces zeros into the first row:

$$P_1 X Q_1 = \begin{pmatrix} X & X & 0 & 0 \\ 0 & X & X & X \\ 0 & X & X & X \\ 0 & X & X & X \\ 0 & X & X & X \\ 0 & X & X & X \end{pmatrix}.$$

This second transformation does not touch the first column, so that the zeros
already introduced are undisturbed. A third transformation introduces zeros
into the second column:

$$P_2 P_1 X Q_1 = \begin{pmatrix} X & X & 0 & 0 \\ 0 & X & X & X \\ 0 & 0 & X & X \\ 0 & 0 & X & X \\ 0 & 0 & X & X \\ 0 & 0 & X & X \end{pmatrix}.$$

Another postmultiplication introduces zeros into the second row:

$$P_2 P_1 X Q_1 Q_2 = \begin{pmatrix} X & X & 0 & 0 \\ 0 & X & X & 0 \\ 0 & 0 & X & X \\ 0 & 0 & X & X \\ 0 & 0 & X & X \\ 0 & 0 & X & X \end{pmatrix}.$$

When $n > k$ the process ends with two premultiplications:

$$P_3 P_2 P_1 X Q_1 Q_2 = \begin{pmatrix} X & X & 0 & 0 \\ 0 & X & X & 0 \\ 0 & 0 & X & X \\ 0 & 0 & 0 & X \\ 0 & 0 & 0 & X \\ 0 & 0 & 0 & X \end{pmatrix}, \qquad P_4 P_3 P_2 P_1 X Q_1 Q_2 = \begin{pmatrix} X & X & 0 & 0 \\ 0 & X & X & 0 \\ 0 & 0 & X & X \\ 0 & 0 & 0 & X \\ 0 & 0 & 0 & 0 \\ 0 & 0 & 0 & 0 \end{pmatrix}.$$

Note for later reference that the product of the $Q$'s has the form

$$Q_1 Q_2 \cdots Q_{k-2} = \begin{pmatrix} 1 & 0 \\ 0 & Q_* \end{pmatrix}. \tag{18.2}$$

## Plane rotations

12. We are going to describe an implicitly shifted QR method for reducing a bidiagonal matrix to diagonal form. Before we describe the algorithm itself, we must digress to introduce the transformations we will use. They are called plane rotations.

13. A *plane (or Givens) rotation* is a matrix of the form

$$P = \begin{pmatrix} c & s \\ -s & c \end{pmatrix},$$

where

$$c^2 + s^2 = 1.$$

The matrix $P$ is clearly orthogonal. Moreover, if the 2-vector $(x \ y)$ is nonzero and we set

$$c = \frac{x}{\sqrt{x^2 + y^2}} \quad \text{and} \quad s = \frac{y}{\sqrt{x^2 + y^2}},$$

then

$$\begin{pmatrix} c & s \\ -s & c \end{pmatrix} \begin{pmatrix} x \\ y \end{pmatrix} = \begin{pmatrix} \sqrt{x^2 + y^2} \\ 0 \end{pmatrix}.$$

14. Plane rotations, suitably embedded in an identity matrix, can be used to introduce zeros in vectors and matrices. For example, with the above definitions of $c$ and $s$, we have

$$\begin{pmatrix} 1 & 0 & 0 & 0 & 0 & 0 \\ 0 & c & 0 & 0 & s & 0 \\ 0 & 0 & 1 & 0 & 0 & 0 \\ 0 & 0 & 0 & 1 & 0 & 0 \\ 0 & -s & 0 & 0 & c & 0 \\ 0 & 0 & 0 & 0 & 0 & 1 \end{pmatrix} \begin{pmatrix} z \\ x \\ z \\ z \\ y \\ z \end{pmatrix} = \begin{pmatrix} z \\ \sqrt{x^2 + y^2} \\ z \\ z \\ 0 \\ z \end{pmatrix}.$$

The plane rotation illustrated above is called a rotation in the $(2,5)$-plane.

15. Some care must be taken in generating plane rotations to avoid overflows and harmful underflows. The following code does the job.

```
1.   rotgen(x, y, c, s, r)
2.       σ = |x| + |y|
3.       if (σ = 0)
4.           c = 1; s = 0; return
5.       end if
6.       r = σ√((x/σ)² + (y/σ)²)
7.       c = x/r; s = y/r; return
8.   end rotgen
```

It is a useful exercise to see what the algorithm does with $x$'s and $y$'s whose squares overflow or underflow.

## The implicit QR algorithm for singular values

16. Suppose we have bidiagonalized our original matrix $A$ by the algorithm described §18.11 to give a bidiagonal matrix $B$. We are now going to describe an iterative algorithm to reduce $B$ to diagonal form by orthogonal transformations. The diagonal entries will be the singular values of $A$. If, in addition, we accumulate the transformations used in the bidiagonalization and the iteration we will get the singular vectors of $A$.

17. We may assume without loss of generality that $B$ is square (the last $n - k$ rows are already zero). Then the matrix

$$C = B^{\mathrm{T}} B$$

is tridiagonal, and its eigenvalues are the squares of the singular values of $B$. For the reasons cited in §§18.7–18.8, we do not want to work with $C$. We will now show how we can achieve the effect of a shifted QR step on $C$ by manipulating $B$.

18. Let us begin by briefly describing the implicitly shifted QR algorithm applied to the tridiagonal matrix $C$.[25] The implicitly shifted QR algorithm with shift $\kappa$ would begin by computing the first column of $C - \kappa I$:

$$\begin{pmatrix} c_{11} - \kappa \\ c_{21} \\ 0 \end{pmatrix}$$

and then reducing it to a multiple $\mathbf{e}_1$. Since the first column of $C$ has only two nonzero elements, it can be reduced by a rotation $Q_0$ in the $(1,2)$-plane.

---

[25] This is an important algorithm in its own right, since the result of reducing a Hermitian matrix to Hessenberg form is a tridiagonal matrix.

We then form $Q_0^{\mathrm{T}} C Q_0$ and determine an orthogonal transformation $Q$ such that $Q^{\mathrm{T}} Q_0^{\mathrm{T}} C Q_0 Q$. If we use the natural variant of the algorithm (16.2), the transformation $Q_0 Q$ will have the same first column as $Q_0$ and hence is the transformation that we would get from an explicit QR step.

19.  Now suppose that we instead postmultiply the transformation $Q_0$ into $B$ and then reduce $B$ to bidiagonal form:

$$\hat{B} = P B Q_0 Q.$$

Then

$$\hat{B}^{\mathrm{T}} \hat{B} = (Q_0 Q)^{\mathrm{T}} C (Q_0 Q)$$

is tridiagonal. Moreover, by (18.2) the first column of $Q_0 Q$ is the same as the first column of $Q_0$. Hence the transformation $Q_0 Q$ is the same as the transformation resulting from an explicit QR step applied to $C$ (see §16.15).

20.  If we choose the shifts appropriately the $(n, n-1)$- and $(n-1, n)$-elements of the matrices $C$ will converge cubically to zero. Since $C = B^{\mathrm{T}} B$, we have

$$c_{n-1,n} = c_{n,n-1} = b_{n,n-1} b_{n-1,n-1}.$$

Hence if $c_{n-1,n-1}$ converges to zero and $b_{n-1,n-1}$ remains bounded away from zero, the superdiagonal element $b_{n,n-1}$ will also converge to zero (see §16.12).

21.  Turning to an implementation of this process, we have to form the first column of $C$. This is easily done:

$$C\mathbf{e}_1 = B^{\mathrm{T}} B \mathbf{e}_1 = \begin{pmatrix} b_{11}^2 \\ b_{11} b_{12} \\ 0 \end{pmatrix}.$$

Thus $Q_0$ is a rotation in the $(1, 2)$-plane that introduces a zero into the second component of

$$C\mathbf{e}_1 = \begin{pmatrix} b_{11}^2 - \kappa \\ b_{11} b_{12} \\ 0 \end{pmatrix}.$$

22.  The reduction of $B Q_0$ to bidiagonal form is especially simple. Since $Q_0$ operates only on the first two columns of $B$, the matrix $B Q_0$ has the form

$$B Q_0 = \begin{pmatrix} \mathrm{X} & \mathrm{X} & 0 & 0 \\ \hat{\mathrm{X}} & \mathrm{X} & \mathrm{X} & 0 \\ 0 & 0 & \mathrm{X} & \mathrm{X} \\ 0 & 0 & 0 & \mathrm{X} \end{pmatrix}.$$

The bump (with a hat) in the $(2,1)$-position can be annihilated by a rotation $P_1$ in the $(1,2)$-plane. The application of $P_1$ introduce a bump in the $(1,3)$-position:

$$P_1 B Q_0 = \begin{pmatrix} X & X & \hat{X} & 0 \\ 0 & X & X & 0 \\ 0 & 0 & X & X \\ 0 & 0 & 0 & X \end{pmatrix}.$$

This bump can be eliminated by postmultiplying by a rotation $Q_1$ in the $(2,3)$-plane — at the cost of introducing a bump in the $(3,2)$-position:

$$P_1 B Q_0 Q_1 = \begin{pmatrix} X & X & 0 & 0 \\ 0 & X & X & 0 \\ 0 & \hat{X} & X & X \\ 0 & 0 & 0 & X \end{pmatrix}.$$

The reduction continues with alternate postmultiplications and premultiplications by plane rotations until the bump is chased out of the matrix:

$$P_2 P_1 B Q_0 Q_1 = \begin{pmatrix} X & X & 0 & 0 \\ 0 & X & X & \hat{X} \\ 0 & 0 & X & X \\ 0 & 0 & 0 & X \end{pmatrix}, \qquad P_2 P_1 B Q_0 Q_1 Q_2 = \begin{pmatrix} X & X & 0 & 0 \\ 0 & X & X & 0 \\ 0 & 0 & X & X \\ 0 & 0 & \hat{X} & X \end{pmatrix},$$

$$P_3 P_2 P_1 B Q_0 Q_1 Q_2 = \begin{pmatrix} X & X & 0 & 0 \\ 0 & X & X & 0 \\ 0 & 0 & X & X \\ 0 & 0 & 0 & X \end{pmatrix}.$$

23.   There are a number of natural candidates for the shift. The traditional shift is the smallest singular value of the trailing $2 \times 2$ submatrix of $B$. With this shift, the superdiagonal element $b_{k-1,k}$ will generally converge cubically to zero. After it has converged, the problem can be deflated by working with the leading principal submatrix of order one less. Singular vectors can be computed by accumulating the transformations applied in the reduction to bidiagonal form and the subsequent QR steps.

- Krylov Sequence Methods

# Krylov Sequence Methods

## Introduction

In the next several lectures we are going to consider algorithms that — loosely speaking — are based on orthogonalizing a sequence of vectors of the form

$$u, Au, A^2u, \ldots.$$

Such a sequences is called a Krylov sequence. The fundamental operation in forming a Krylov sequence is the computation of the product of $A$ with a vector. For this reason Krylov sequence methods are well suited for large sparse problems in which it is impractical to decompose the matrix $A$.

Unfortunately, Krylov sequence methods are not easy to manage, and the resulting programs are often long and intricate. It has been my experience that these programs cannot be treated as black boxes; you need to know something about the basic algorithm to use them intelligently. This requirement sets the tone of the following lectures. We will present the general algorithm and glide around the implementation details.

We will begin with Krylov methods for the eigenvalue problem. Computing the complete eigensystem of a large, sparse matrix is generally out of the question. The matrix of eigenvectors will seldom be sparse and cannot be stored on a computer. Fortunately, we are often interested in a small part of the eigensystem of the matrix — usually a part corresponding to a subspace called an invariant subspace. Accordingly, we will begin with a treatment of these subspaces.

## Invariant subspaces

1. Let $A \in \mathbf{C}^{n \times n}$ and let $\mathcal{U}$ be a subspace of $\mathbf{C}^n$ of dimension $k$. Then $\mathcal{U}$ is an *invariant subspace* of $A$ if

$$A\mathcal{U} \equiv \{Au : u \in \mathcal{U}\} \subset \mathcal{U}.$$

In other words an invariant subspace of $A$ is unchanged (or possibly diminished) on multiplication by the $A$.

2.  An invariant subspace is in some sense a generalization of the notion of eigenvector. For if $Au = \lambda u$ and $\lambda \neq 0$, the spaces spanned by $u$ and $Au$ are the same.

3.  Let the columns of $U = (u_1 \; \cdots \; u_k)$ form a basis for $\mathcal{U}$. Since $Au_j \in \mathcal{U}$, it must be expressible as a unique linear combination of the columns of $U$; that is, there is a unique vector $h_j$ such that

$$Au_j = Uh_j, \qquad j = 1, \ldots, k.$$

If we set $H = (h_1 \; \cdots \; h_k)$, then

$$AU = UH.$$

The matrix $H$ is a representation of $A$ with respect to the basis $U$. For if $x = Uz$, then

$$Ax = AUz = U(Hz);$$

i.e., if $z$ represents the vector $x$ with respect to the basis $U$, then $Hz$ represents the vector $Ax$ with respect to the same basis.

4.  There is a nice relation between the eigensystem of $H$ and that of $A$. if $Ax = \lambda x$, where $x = Uz$, then

$$UHz = Ax = \lambda x = \lambda Uz;$$

and since the columns of $U$ are independent,

$$Hz = \lambda z.$$

Conversely if $Hz = \lambda z$, then

$$AUz = UHz = \lambda Uz,$$

so that $Uz$ is an eigenvector of $A$. Thus there is a 1-1 correspondence between eigenpairs of $H$ and a subset of the eigenpairs of $A$.

5.  An important property of invariant subspaces is that its members are uncoupled from eigenvectors corresponding to eigenvalues not in the subspace. For suppose $(\lambda, x)$ is an eigenpair of $A$, where $\lambda$ is not an eigenvalue of $H$, and let $(\lambda, y^{\mathrm{H}})$ be the corresponding left eigenpair. Then

$$\lambda y^{\mathrm{H}}U = y^{\mathrm{H}}AU = y^{\mathrm{H}}UH.$$

If $y^{\mathrm{H}}U \neq 0$, then $(\lambda, y^{\mathrm{H}}U)$ is an eigenpair of $H$, contrary to hypothesis. Thus the column of $U$ and their linear combinations — i.e., the members of $\mathcal{U}$ — have no components along $x$ in the sense of §14.19.

6.    If $X$ is nonsingular and we replace $U$ by $UX$, then

$$A(UX) = (UX)(X^{-1}HX);$$

that is, $H$ is replaced by the similar matrix $X^{-1}HX$. Thus we are free to perform arbitrary similarities on $H$. Such similarities can be used to break up invariant subspaces.

7.  Suppose, for example, we can find a matrix $X = (X_1\ X_2)$ such that

$$\hat{H} = X^{-1}HX = (X_1\ X_2)^{-1}H(X_1\ X_2) = \begin{pmatrix} \hat{H}_{11} & \hat{H}_{12} \\ 0 & \hat{H}_{22} \end{pmatrix}.$$

Then partitioning

$$U(X_1\ X_2) = (\hat{U}_1\ \hat{U}_2), \qquad\qquad (19.1)$$

we see that

$$A(\hat{U}_1\ \hat{U}_2) = (\hat{U}_1\ \hat{U}_2) \begin{pmatrix} \hat{H}_{11} & \hat{H}_{12} \\ 0 & \hat{H}_{22} \end{pmatrix}.$$

It follows that $A\hat{U}_1 = \hat{U}_1\hat{H}_{11}$, and the columns of $\hat{U}_1$ span an invariant subspace that is a subspace of the original invariant subspace spanned by $U$.

8.  If we go further and block diagonalize $H$ so that

$$(X_1\ X_2)^{-1}H(X_1\ X_2) = \begin{pmatrix} \hat{H}_{11} & 0 \\ 0 & \hat{H}_{22} \end{pmatrix},$$

then both $\hat{U}_1$ and $\hat{U}_2$ span invariant subspaces.

9.    In practice we will seldom have an exact invariant subspace. Instead we will have

$$AU - UH = G,$$

where $G$ is small. In this case we have the following analogue of the results of §§14.1–14.6. The proof is left as an exercise.

---

Let $U$ have orthonormal columns and let

$$AU - UH = G.$$

Then there is a matrix $E = -GU^H$ with

$$\|E\|_p = \|G\|_p, \qquad p = 2, F$$

such that

$$(A + E)U = UH.$$

Moreover, $\|G\|_F$ assumes its minimum value when

$$H = U^H AU. \qquad\qquad (19.2)$$

---

The matrix $H$ defined by (19.2) is sometimes called the *matrix Rayleigh quotient,* since it is defined in analogy with the scalar Rayleigh quotient. If $(\mu, z)$ is an eigenpair of $H$, the pair $(\mu, Uz)$ is an eigenpair of $A + E$. If the pair is well conditioned and $E$ is small, it is a good approximation to an eigenpair of $A$. The pair $(\mu, Uz)$ is called a *Ritz pair.*

## Krylov subspaces

10. We have already noted that a step of the power method can easily take advantage of sparsity in the matrix, since it involves only a matrix-vector multiplication. However, the convergence of the powers $A^i u$ to the dominant eigenvector can be extremely slow. For example, consider the matrix

$$A = \mathrm{diag}(1, .95, .95^2, \ldots, .95^{99}),$$

whose dominant eigenvector is $\mathbf{e}_1$. The first column of Table 19.1 gives the norm of the vector consisting of the last 99 components of $A^k u / \|A^k u\|_2$, where $u$ is a randomly chosen vector. After 15 iterations we have at best a two-digit approximation to the eigenvector.

11. On the other hand, suppose we try to take advantage of all the powers we have generated. Specifically, let

$$K_k = (u \;\; Au \;\; A^2 u \;\; \cdots \;\; A^{k-1} u)$$

Table 19.1. *Power and Krylov approximations.*

| $k$ | Power | Krylov |
|----|----------|----------|
| 1 | 9.7e−01 | 9.7e−01 |
| 2 | 4.9e−01 | 4.8e−01 |
| 3 | 3.0e−01 | 2.2e−01 |
| 4 | 2.2e−01 | 1.3e−01 |
| 5 | 1.7e−01 | 5.8e−02 |
| 6 | 1.4e−01 | 3.4e−02 |
| 7 | 1.1e−01 | 2.1e−02 |
| 8 | 9.7e−02 | 1.2e−02 |
| 9 | 8.1e−02 | 4.8e−03 |
| 10 | 6.9e−02 | 1.6e−03 |
| 11 | 5.8e−02 | 6.2e−04 |
| 12 | 5.0e−02 | 1.9e−04 |
| 13 | 4.3e−02 | 6.3e−05 |
| 14 | 3.7e−02 | 1.8e−05 |
| 15 | 3.2e−02 | 3.8e−06 |

and let us look for an approximation to the eigenvector $\mathbf{e}_1$ in the form

$$\mathbf{e}_1 =_2 K_k z,$$

where "$=_2$" means approximation in the least squares sense. The third column in Table 19.1 shows the results. The method starts off slowly but converges with increasing rapidity. By the fifteenth iteration it has produced an approx imation accurate to about six figures.

12.  The sequence $u, Au, A^2 u, \ldots$ is called a *Krylov sequence,* and the matrix $K_k$ is called the $k$th Krylov matrix. (For simplicity of notation, we do not express the dependence of $K_k$ on the starting vector $u$.) We have seen that in one case the spaces spanned by the Krylov matrices have increasingly good approximations to the dominant eigenvector. It turns out that it also has good approximations to other eigenvectors. Consequently, it will be worth our while to investigate techniques for extracting these vectors.

## Arnoldi decompositions

13.  One problem with the Krylov sequence is that, in general, its vectors tend toward the dominant eigenvector, and hence the Krylov matrices become increasingly ill conditioned. We can solve this problem by orthogonalizing the Krylov vectors. Specifically, having found an orthonormal basis

$$U_k = (u_1 \ u_2 \ \cdots \ u_k)$$

for the column space of $K_k$ we can extend it by applying one step of the Gram–Schmidt algorithm to $A^k u$. Now the vector $Au_k$ is a linear combination of $Au, A^2 u, \ldots, A^k u$. Hence we can obtain the same effect by orthogonalizing $Au_k$. Specifically, we compute

$$h_{ik} = u_i^{\mathrm{H}} Au_k, \qquad i = 1, \ldots, j$$

and then

$$h_{k+1,k} u_{k+1} = Au_k - \sum_{i=1}^{k} h_{ik} u_k,$$

where $h_{k+1,k}$ is chosen to normalize $u_{k+1}$. This process can be continued as long as the vectors in the Krylov sequence remain linearly independent.

14.  The above process is identical to the Arnoldi reduction of §16.15, except that we stop after a finite number of steps.[26]  The results can be written

---

[26] Warning: Some care must be taken to insure that the vectors $u_k$ remain orthogonal to working accuracy in the presence of rounding error. The usual technique is called reorthogonalization and is discussed in Åke Björck's *Numerical Methods for Least Squares Problems,* SIAM, 1996.

succinctly in terms of matrices. Let

$$
H_k = \begin{pmatrix}
h_{11} & h_{12} & h_{13} & \cdots & h_{1,k-1} & h_{1k} \\
h_{21} & h_{22} & h_{23} & \cdots & h_{2,k-1} & h_{2k} \\
0 & h_{32} & h_{33} & \cdots & h_{3,k-1} & h_{3k} \\
0 & 0 & h_{43} & \cdots & h_{4,k-1} & h_{4k} \\
\vdots & \vdots & \vdots & & \vdots & \vdots \\
0 & 0 & 0 & \cdots & h_{k,k-1} & h_{kk}
\end{pmatrix}.
$$

Then $H_k$ is upper Hessenberg, and

$$
AU_k = U_k H_k + h_{k+1,k} u_{k+1} \mathbf{e}_k^{\mathrm{T}}. \tag{19.3}
$$

Any decomposition of the form (19.3), where $H_k$ is Hessenberg, $U_k^{\mathrm{H}} U_k = I$, and $U_k^{\mathrm{H}} u_{k+1} = 0$, is called an *Arnoldi decomposition of length $k$*. If $H$ is unreduced and $h_{k+1,k} \neq 0$, the decomposition is uniquely determined by the starting vector $u$.

15. Since $U_k^{\mathrm{H}} u_{k+1} = 0$, we have

$$
H_k = U_k^{\mathrm{H}} AU_k.
$$

Thus in the terminology of §19.9, $H_k$ is the Rayleigh quotient corresponding to $U_k$, and its eigenpairs are Ritz pairs.

16.  Let $(\mu, z)$ be an eigenpair of $H_k$ with $\|z\|_2 = 1$. Then

$$
AU_k z - \mu U_k z = h_{k+1,k} \zeta_k u_{k+1},
$$

where $\zeta_k$ is the last component of $z$. Thus by the result of §14.3 $(\mu, U_k z)$ is an exact eigenpair of $A + E$, where $\|E\|_{\mathrm{F}} = |h_{k+1,k} \zeta_k|$. This gives us a criterion for accepting the pair $(\mu, U_k z)$ as an eigenpair of $A$.

17.  We now have a potentially useful algorithm — generally called *Arnoldi's method*. Given a starting vector $u$ we perform the following operations.

1.   Generate the Arnoldi decomposition of length $k$
2.   Compute the Ritz pairs and decide which are acceptable
3.   If you have not found what you need, increase $k$ and repeat

The algorithm has two nice aspects. First, the matrix $H_k$ is already in Hessenberg form, so that we can immediately apply the QR algorithm to find its eigenvalues. Second, after we increase $k$, say to $k + p$, we only have to orthogonalize $p$ vectors to compute the $(k + p)$th Arnoldi decomposition. The work we have done previously is not thrown away.

Unfortunately, the algorithm has its drawbacks. In the first place, if $A$ is large we cannot increase $k$ indefinitely, since $U_k$ requires $n \times k$ memory locations to store. Moreover, we have little control over which eigenpairs the algorithm finds. In a given application we will be interested in a certain set of eigenpairs — for example, eigenvalues lying near the imaginary axis. There is nothing in the algorithm to force desired eigenvectors into the subspace or to discard undesired ones. We will now show how to implicitly restart the algorithm with a new Arnoldi decomposition in which (in exact arithmetic) the unwanted eigenvalues have been purged from $H$.

## Implicit restarting

18. We begin by asking how to cast an undesired eigenvalue out of an unreduced Hessenberg matrix $H$. Let $\mu$ be the eigenvalue, and suppose we perform one step of the QR algorithm with shift $\mu$. The first step is to determine an orthogonal matrix $Q$ such that

$$R = Q^{\mathrm{H}}(H - \mu I)$$

is upper triangular. Now $H - \mu I$ is singular, and hence $R$ must have a zero on its diagonal. Because $H$ was unreduced, that zero cannot occur at a diagonal position other than the last (see the reduction in §15.26). Consequently, the last row of $R$ is zero and so is the last row of $RQ$. Hence $\hat{H} = RQ + \mu I$ has the form

$$\begin{pmatrix} \hat{H}_* & \hat{h} \\ 0 & \mu \end{pmatrix}.$$

In other words the shifted QR step has found the eigenvalue exactly and has deflated the problem.

In the presence of rounding error we cannot expect the last element of $R$ to be nonzero. This means that the matrix $\hat{H}$ will have the form

$$\begin{pmatrix} \hat{H}_* & \hat{h} \\ \hat{h}_{k,k-1}\mathbf{e}_{k-1}^{\mathrm{T}} & \hat{\mu} \end{pmatrix}.$$

19. We are now going to show how to use the transformation $Q$ to reduce the size of the Arnoldi decomposition. The first step is to note that $Q = P_{12}P_{23}\cdots P_{n-1,n}$, where $P_{i,i+1}$ is a rotation in the $(i, i+1)$-plane. Consequently, $Q$ is Hessenberg and can be partitioned in the form

$$Q = \begin{pmatrix} Q_* & q \\ q_{k,k-1}\mathbf{e}_{k-1}^{\mathrm{T}} & q_{k,k} \end{pmatrix}.$$

From the relation

$$AU_k = U_k H_k + h_{k+1,k}u_{k+1}\mathbf{e}_k^{\mathrm{T}},$$

we have

$$AU_k Q = U_k QQ^{\mathrm{H}}H_k Q + h_{k+1,k}u_{k+1}\mathbf{e}_k^{\mathrm{T}}Q.$$

If we partition

$$\hat{U}_k = U_k Q = (\hat{U}_{k-1}\ \hat{u}_k),$$

then

$$A(\hat{U}_{k-1}\ \hat{u}_k) = (\hat{U}_{k-1}\ \hat{u}_k)\begin{pmatrix} \hat{H}_* & \hat{h} \\ \hat{h}_{k,k-1}\mathbf{e}_{k-1}^{\mathrm{T}} & \hat{\mu} \end{pmatrix} + h_{k+1,k}u_{k+1}(q_{k,k-1}\mathbf{e}_{k-1}^{\mathrm{T}}\ q_{k,k}).$$

Computing the first $k-1$ columns of this partition, we get

$$A\hat{U}_{k-1} = \hat{U}_{k-1}\hat{H}_* + (\hat{h}_{k,k-1}\hat{u}_k + h_{k+1,k}q_{k,k-1}u_{k+1})\mathbf{e}_{k-1}^{\mathrm{T}}. \qquad (19.4)$$

The matrix $\hat{H}_*$ is Hessenberg. The vector $\hat{h}_{k,k-1}\hat{u}_k + h_{k+1,k}q_{k,k-1}u_{k+1}$ is orthogonal to the columns of $\hat{U}_{k-1}$. Hence (19.4) is an Arnoldi decomposition of length $k-1$. With exact computations, the eigenvalue $\mu$ is not present in $H_*$.

20. The process may be repeated to remove other unwanted values from $H$. If the matrix is real, complex eigenvalues can be removed two at a time via an implicit double shift. The key observation here is that $Q$ is zero below its second subdiagonal element, so that truncating the last two columns and adjusting the residual results in an Arnoldi decomposition. Once undesired eigenvalues have been removed from $H$, the Arnoldi decomposition may be expanded to one of order $k$ and the process repeated.

## Deflation

21. There are two important additions to the algorithm that are beyond the scope of these lectures. They correspond to the two kinds of deflation in §19.7 and §19.8. They are used for quite distinct purposes.

22. First, as Ritz pairs converge they can be locked into the decomposition. The procedure amounts to computing an Arnoldi decomposition of the form

$$A(U_1\ U_2) = (U_1\ U_2)\begin{pmatrix} H_{11} & H_{12} \\ 0 & H_{21} \end{pmatrix} + h_{k+1,k}u_{k+1}\mathbf{e}_k^{\mathrm{T}}.$$

When this is done, one can work with the part of the decomposition corresponding to $U_2$, thus saving multiplications by $A$. (However, care must be taken to maintain orthogonality to the columns of $U_1$.)

23. The second addition concerns unwanted Ritz pairs. The restarting procedure will tend to purge the unwanted eigenvalues from $H$. But the columns of $U$ may have significant components (see §14.19) along the eigenvectors corresponding to the purged pairs, which will then reappear as the decomposition

is expanded. If certain pairs are too persistent, it is best to keep them around
by computing a block diagonal decomposition of the form

$$A(U_1 \ U_2) = (U_1 \ U_2) \begin{pmatrix} H_{11} & 0 \\ 0 & H_{22} \end{pmatrix} + \eta u_{k+1} \mathbf{e}_k^{\mathrm{T}},$$

where $H_{11}$ contains the unwanted eigenvalues. This insures that $U_2$ has negligible components along the unwanted eigenvectors. We can then compute an
Arnoldi decomposition by reorthogonalizing the relation

$$AU_2 = H_{22}U_2 + \eta u_{k+1} \mathbf{e}_m^{\mathrm{T}},$$

where $m$ is the order of $H_2$.

# Krylov Sequence Methods

The Lanczos Algorithm
Relation to Orthogonal Polynomials
Golub–Kahan–Lanczos Bidiagonalization

## The Lanczos algorithm

1. Let $A$ be symmetric and let

$$AU_k = U_k H_k + \eta_k u_{k+1} \mathbf{e}_k^{\mathrm{T}}$$

be an Arnoldi decomposition. Then

$$H_k = U_k^{\mathrm{T}} A U_k$$

is symmetric. Since $H_k$ is upper Hessenberg, it is also lower Hessenberg and hence tridiagonal. To stress this fact we will write the symmetric decomposition in the form

$$AU_k = UT_k + \eta_k u_{k+1} \mathbf{e}_k^{\mathrm{T}}, \tag{20.1}$$

where

$$T_k = \begin{pmatrix} \alpha_1 & \beta_1 & & & & \\ \beta_1 & \alpha_2 & \beta_2 & & & \\ & \beta_2 & \alpha_3 & \beta_3 & & \\ & & \ddots & \ddots & \ddots & \\ & & & \beta_{k-2} & \alpha_{k-1} & \beta_{k-1} \\ & & & & \beta_{k-1} & \alpha_k \end{pmatrix}.$$

2. The tridiagonality of $T_k$ causes the algorithm for generating the decomposition to simplify. Partitioning the matrices in (20.1), we get

$$A(u_1\ u_2\ u_3\ \cdots) = (u_1\ u_2\ u_3\ \cdots) \begin{pmatrix} \alpha_1 & \beta_1 & & \\ \beta_1 & \alpha_2 & \beta_2 & \\ & \beta_2 & \alpha_3 & \beta_3 \\ & & \ddots & \ddots & \ddots \end{pmatrix} + \eta_k u_{k+1} \mathbf{e}_k^{\mathrm{T}}.$$

It follows that

$$\beta_1 u_2 = Au_1 - \alpha_1 u_1,$$
$$\beta_2 u_3 = Au_2 - \alpha_2 u_2 - \beta_1 u_1,$$

and in general

$$\beta_k u_{k+1} = Au_k - \alpha_k u_k - \beta_{k-1} u_{k-1}. \tag{20.2}$$

177

3. It is easy to generate the coefficients $\alpha_k$ and $\beta_k$. From the relation $T_k = U_k^T U_k$, we get

$$\alpha_k = u_k A u_k$$

and

$$\beta_{k-1} = u_{k-1}^T A u_k.$$

Alternatively, we can choose $\beta_k$ so that $u_{k+1}$ in (20.2) is normalized. Although the two choices are mathematically equivalent, numerically the second choice is better. The leads to the *Lanczos algorithm* below.[27]

1.  $u_0 = 0$
2.  $\beta_0 = 1/\|u\|_2$; $u_1 = \beta_0\|u\|$
3.  **for** $k = 1, 2, \ldots$
4.      $u_{k+1} = Au_k$
5.      $\alpha_k = u_k^T u_{k+1}$
6.      $u_{k+1} = u_{k+1} - \alpha_k u_k - \beta_{k-1} u_{k-1}$
7.      $\beta_k = 1/\|u_{k+1}\|_2$
8.      $u_{k+1} = \beta_k u_{k+1}$
9.  **end for** $k$

4. The advantage of this algorithm is that it is never necessary to have more than three of the $u_k$ at a time in memory. Moreover, the orthogonalization process is much cheaper than for the general Arnoldi method, requiring only $O(n)$ operations per vector.

   At some point one can stop the orthogonalization process and compute eigenpairs $(\mu, z)$ from $T_k$. If they are satisfactory, we can generate the corresponding eigenvectors of $A$ in the form $U_k z$. The columns of $U_k$ may be regenerated by the Lanczos process; or if we have squirreled them away on a backing store, we can read them in again.

5. Unfortunately, this naive procedure does not work. The problem is loss of orthogonality in the vectors $u_k$ due to rounding error. To see why this loss is a problem, suppose that for some $k$ there is an accurate Ritz pair $(\mu, U_k z)$. If the vectors $u$ are strictly orthogonal, no second approximation of $U_k z$ can appear in the spaces spanned by $U_{k+j}$. However, if orthogonality is lost, multiple copies can appear. Thus the Lanczos process, in the presence of rounding error, can produce the same Ritz pairs over and over.

6. The cure for the difficulty is simple in conception. When a satisfactory Ritz pair is found, the Ritz vector is kept around, and the $u_k$ are orthogonalized against it. Thus the work increases as we find more Ritz pairs, but the process is still much less expensive than the general Arnoldi process.

---

[27]The name is pronounced Lantzosh with the accent on the first syllable. I am told it means lancer.

7. This modified process is called the Lanczos algorithm with selective re-orthogonalization. It is the method of choice for large, sparse symmetric eigenvalue problems. But it is very tricky to implement, and the person who needs it should go to an archive like Netlib to get code.

## Relation to orthogonal polynomials

8. The fact that orthogonal polynomials and the vectors in the Lanczos process satisfy a three-term recurrence suggests that there might be some relation between the two. To see what that relation might be, let

$$\mathcal{K}_k = \mathrm{span}(u, Au, A^2u, \ldots, A^{k-1}u)$$

be the $k$th *Krylov subspace* associated with $A$ and $u$. We will suppose that

$$\dim(\mathcal{K}_k) = k;$$

i.e., the Krylov vectors that span $\mathcal{K}_k$ are linearly independent.

If $v \in \mathcal{K}_k$, then

$$v = c_0 u + c_1 Au + \cdots + c_{k-1} A^{k-1} u.$$

Consequently, if we set

$$p(x) = c_0 + c_1 x + \cdots + c_{k-1} x^{k-1},$$

we have

$$v = p(A)u.$$

Thus there is a 1-1 correspondence between polynomials of degree $k-1$ or less and members of $\mathcal{K}_k$.

9. Now let us define an inner product between two polynomials by

$$(p, q) = [p(A)u]^{\mathrm{T}}[q(A)u].$$

To see that this formula defines an inner product, let $p \neq 0$. Then by the independence of the Krylov vectors, $p(A)u \neq 0$, and

$$(p, p) = \|p(A)u\|_2^2 > 0.$$

Thus our inner product is definite. It is easy to show that it is bilinear. It is symmetric because

$$(p, q) = (p, q)^{\mathrm{T}} = [u^{\mathrm{T}} p(A)^{\mathrm{T}} q(A)u]^{\mathrm{T}} = u^{\mathrm{T}} q(A)^{\mathrm{T}} p(A)u = (q, p).$$

10. The inner product $(\cdot,\cdot)$ on the space of polynomials generates a sequence $p_0, p_1, \ldots$ of orthogonal polynomials (see the comments in §7.17), normalized so that $(p_j, p_j) = 1$. If we define $u_j = p_j(A)u$, then

$$u_i^{\mathrm{T}} u_j = (p_i, p_j),$$

and hence the vectors $u_j$ are orthonormal.

However, if $A$ is not symmetric these polynomials do not satisfy a three-term recurrence. The reason is that to derive the three-term recurrence we needed the fact that $(tp, q) = (p, tq)$. For symmetric matrices we have

$$(tp, q) = u^{\mathrm{T}}[p(A)^{\mathrm{T}} A^{\mathrm{T}}]q(A)u = u^{\mathrm{T}} p(A)^{\mathrm{T}}[Aq(A)]u = (q, tp).$$

Thus when the Arnoldi process is applied to a symmetric matrix, the result is a tridiagonal matrix whose elements are the coefficients of the three-term recurrence for a sequence of orthogonal polynomials.

## Golub–Kahan–Lanczos bidiagonalization

11. We have seen that the symmetric eigenvalue problem and the singular value decomposition are closely related — the singular values of a square matrix $X$ are the square roots of the eigenvalues of $X^{\mathrm{T}}X$. Consequently, we can calculate singular values by applying the Lanczos algorithm to $X^{\mathrm{T}}X$. The matrix-vector product required by the algorithm can be computed in the form $X^{\mathrm{T}}(Xu)$ — i.e., a multiplication by $X$ followed by a multiplication by $X^{\mathrm{T}}$.

12. As a general rule singular value problems are best treated on their own terms and not reduced to an eigenproblem (see §§18.5–18.9). Now the Lanczos algorithm can be regarded as using a Krylov sequence to tridiagonalize $X^{\mathrm{T}}X$. The natural analogue is to bidiagonalize $X$. Since the bidiagonalization algorithm described in §18.11 involves two-sided transformations, the corresponding Lanczos-like process will involve two Krylov sequence. It is called *Golub–Kahan–Lanczos bidiagonalization*.

13. For simplicity we will assume that $X$ is square. The algorithm can be derived directly from the bidiagonalization algorithm in §18.11. Specifically, let $v_1$ be given with $\|v_1\|_2 = 1$, and let $Q_0$ be an orthogonal matrix whose first column is $v_1$. If we apply the bidiagonalization algorithm to $XQ_0$, we end up with orthogonal matrices $U$ and $Q$ such that $U^{\mathrm{T}}(XQ_0)Q$ is bidiagonal. Moreover, $Q$ has the form

$$\begin{pmatrix} 1 & 0 \\ 0 & Q_* \end{pmatrix}$$

[see (18.2)]. Hence if we set $V = Q_0 Q$, the first column of $V$ will be $v_1$.

We have thus shown that given any vector $v_1$ with $\|v_1\|_2 = 1$, there are orthogonal matrices $U$ and $V$ such that

$$V \mathbf{e}_1 = v_1$$

and

$$
U^{\mathrm{T}} X V = \begin{pmatrix} \alpha_1 & \beta_1 & & & & \\ & \alpha_2 & \beta_2 & & & \\ & & \alpha_3 & \beta_3 & & \\ & & & \ddots & \ddots & \\ & & & & \alpha_{n-1} & \beta_{n-1} \\ & & & & & \alpha_n \end{pmatrix}.
\tag{20.3}
$$

The constants $\alpha_k$ and $\beta_k$ are given by

$$
\alpha_k = u_k^{\mathrm{T}} X v_k \quad \text{and} \quad \beta_k = u_k^{\mathrm{T}} X v_{k+1}.
\tag{20.4}
$$

14. From the bidiagonal form (20.3) we may derive a double recursion for the columns $u_k$ and $v_k$ of $U$ and $V$. Multiplying by $U$, we have

$$
X(v_1 \; v_2 \; v_3 \; \cdots) = (u_1 \; u_2 \; u_3 \; \cdots) \begin{pmatrix} \alpha_1 & \beta_1 & & \\ & \alpha_2 & \beta_2 & \\ & & \alpha_3 & \beta_3 \\ & & & \ddots & \ddots \end{pmatrix}.
$$

On computing the $k$th column of this relation, we find that $Xv_k = \beta_{k-1}u_{k-1} + \alpha_k v_k$ or

$$
\alpha_k u_k = X v_k - \beta_{k-1} u_{k-1}.
$$

(To get us started we assume that $\beta_0 = 0$.) From the relation

$$
X^{\mathrm{T}}(u_1 \; u_2 \; u_3 \; \cdots) = (v_1 \; v_2 \; v_3 \; \cdots) \begin{pmatrix} \alpha_1 & & & \\ \beta_1 & \alpha_2 & & \\ & \beta_2 & \alpha_3 & \\ & & \ddots & \ddots \end{pmatrix}
$$

we get $X^{\mathrm{T}} u_k = \alpha_k v_k + \beta_k v_{k+1}$ or

$$
\beta_k v_{k+1} = X^{\mathrm{T}} u_k - \alpha_k v_k.
$$

15. We do not quite have a recursion because the formulas (20.4) for the $\alpha_k$ and $\beta_k$ involve the very vectors we are trying to compute. But since the columns of $U$ and $V$ are normalized, we must have

$$
\alpha_k = \| X v_k - \beta_{k-1} u_{k-1} \|_2 \quad \text{and} \quad \beta_k = \| X^{\mathrm{T}} u_k - \alpha_k v_k \|_2.
$$

We summarize the recursion in the following algorithm.

1.   $\beta_0 = 0$
2.   **for** $k = 1, 2, \ldots$
3.        $u_k = X v_k - \beta_{k-1} u_{k-1}$
4.        $\alpha_k = \|u_k\|_2$
5.        $u_k = u_k / \alpha_k$
6.        $v_k = X^{\mathrm{T}} u_k - \alpha_k v_k$
7.        $\beta_k = \|v_{k+1}\|_2$
8.        $v_{k+1} = v_{k+1} / \beta_k$
9.   **end for** $k$

16. Although Golub–Kahan–Lanczos bidiagonalization can be used to find singular values, it is more important as the basis for iterative algorithms to solve least squares problems and nonsymmetric linear systems. It would take us too far afield to develop these algorithms. Instead, in the following lectures we will treat a method for solving positive definite systems that is closely related to the Lanczos algorithm.

# Krylov Sequence Methods

Linear Systems, Errors, and Residuals
Descending to a Solution
Conjugate Directions
The Method of Conjugate Gradients
Termination

## Linear systems, errors, and residuals

1. We now turn to the iterative solution of large linear systems. Until further notice $A$ will be a positive definite matrix of order $n$, where $n$ is presumed large. As in the large eigenvalue problem we will suppose that the product of $A$ with a vector can be calculated but that it is impractical to manipulate $A$ itself. Our immediate goal is to derive the method of conjugate gradients for positive definite systems.

2. Consider the linear system

$$Ax_* = b. \tag{21.1}$$

For any vector $x$ there are two ways of measuring the quality of $x$ as an approximate solution of (21.1): the *error*

$$e = x - x_*$$

and the *residual*

$$r = b - Ax.$$

The two quantities are related by the equations

$$r = -Ae \quad \text{and} \quad e = -A^{-1}r. \tag{21.2}$$

3. In practice, the error is not computable; for if we knew its value we could find $x_*$ in the form $x_* = x + e$. On the other hand, the computation of the residual involves only the subtraction of the product $Ax$ from $b$. This is one reason why the residual features prominently in algorithms for solving large linear systems.

## Descending to a solution

4. It is possible — in fact easy — to write down a quadratic function whose minimum is $x_*$. Specifically, let

$$\varphi(x) = \frac{1}{2}e^{\mathrm{T}}Ae.$$

Since $A$ is positive definite, $\varphi(x) \geq 0$. Moreover, $\varphi(x) = 0$ if and only if $e = 0$, or equivalently if $x = x_*$.

5.  Since the minimum of $\varphi(x)$ is $x_*$, we can solve the system $Ax_* = b$ by minimizing $\varphi$. We will adopt the following strategy. Given a starting point $x_1$, choose a direction $s_1$ and determine $\alpha_1$ so that

$$\varphi(x_1 + \alpha_1 s_1) = \min_\alpha \varphi(x_1 + \alpha s_1).$$

Now $\varphi(x_1 + \alpha s_1)$ is a quadratic function of $\alpha$ with positive second derivative. Hence a minimizing $\alpha_1$ exists and is unique. This process advances us to the point

$$x_2 = x_1 + \alpha_1 s_1.$$

In general, given $x_k$, choose a direction $s_k$, determine $\alpha_k$ to minimize $\varphi(x_k + \alpha s_k)$, and set

$$x_{k+1} = x_k + \alpha_k s_k. \tag{21.3}$$

The residuals for this process obviously satisfy the recurrence

$$r_{k+1} = r_k - \alpha_k A s_k. \tag{21.4}$$

6.  It can be shown that if the $s_k$ are properly chosen (for example, we might repeatedly cycle through the unit vectors $e_1, \ldots, e_n$), the $x_k$ will converge to $x_*$. The fly in the ointment is that we cannot evaluate $\varphi$, which is defined in terms of the error. It is true that by using (21.2) we can express $\varphi$ in terms of the residual in the form

$$\varphi(x) = r^{\mathrm{T}} A^{-1} r, \tag{21.5}$$

but we don't know $A^{-1}$ either. What saves our program is that we can recognize when $\varphi(x + \alpha s)$ is minimal without actually evaluating $\varphi$.

7.  It will be convenient to cast the problem in more general terms. Let

$$S_k = (s_1 \; s_2 \; \cdots \; s_k)$$

consist of linearly independent directions $s_j$ and let

$$\hat{x}_{k+1} = x_1 + S_k \hat{a}_k = x_1 + \hat{\alpha}_1 s_1 + \cdots + \hat{\alpha}_k s_k.$$

Then we can pose the problem of determining $\hat{a}_k$ so that

$$\varphi(x_1 + S_k \hat{a}_k) = \min.$$

Note that $\hat{x}_k$ will in general give a smaller value of $\varphi$ than $x_k$ from the sequential minimization (21.3).

8.  The following result characterizes the point $\hat{x}_k$ at which the minimum occurs.

> The minimum of the function $\varphi(x_1 + S_k \hat{a})$ occurs at the unique point $\hat{x}_{k+1} = x_1 + S_k \hat{a}_k$ for which
>
> $$S_k^{\mathrm{T}} \hat{r}_k = 0.$$

To see this, write in analogy with (21.4)

$$r = r_1 - ASa.$$

Then by (21.5),

$$\varphi(x_1 + S_k \hat{a}) = (r_1 - AS_k \hat{a})^{\mathrm{T}} A^{-1} (r_1 - AS_k \hat{a}) = \|A^{-\frac{1}{2}} r_1 - A^{\frac{1}{2}} S_k \hat{a}\|_2^2,$$

where $A^{\frac{1}{2}}$ is the positive definite square root of $A$. This equation exhibits our minimization problem as a least squares problem with least squares matrix $A^{\frac{1}{2}} S_k$. Since the $s_j$ are linearly independent, $A^{\frac{1}{2}} S_k$ is of full rank, and the least squares problem has a unique solution $\hat{a}_k$ satisfying the normal equations:

$$S_k^{\mathrm{T}} A S_k \hat{a}_k = S_k^{\mathrm{T}} r_1. \tag{21.6}$$

Equivalently,

$$0 = S_k^{\mathrm{T}} (r_1 - AS_k \hat{a}_k) = S_k^{\mathrm{T}} \hat{r},$$

which is what we wanted to show.

9. In plain words the result says that at the minimum the residual is orthogonal to the space spanned by the directions. For our original problem of sequential minimization, it follows from the recurrence $r_{k+1} = r_k - \alpha_k A s_k$ that

$$0 = s_k^{\mathrm{T}} r_{k+1} = s_k^{\mathrm{T}} r_k - \alpha_k s_k^{\mathrm{T}} A s_k$$

or

$$\alpha_k = \frac{s_k^{\mathrm{T}} r_k}{s_k^{\mathrm{T}} A s_k}. \tag{21.7}$$

This easily calculated expression is what makes our program of sequential minimization feasible.

## Conjugate directions

10. Given $k$ directions $s_1, \ldots, s_k$, we have posed two distinct ways of approximating the solution to $Ax = b$. The first is the method of sequential minimization, which produces the approximation $x_{k+1}$. The second is the result $\hat{x}_{k+1}$ of minimizing $\varphi(x_1 + S_k \hat{a})$. The advantage of the approximate solution $x_{k+1}$ is that it is easy to compute. Moreover, once $s_j$ has been used, it can be discarded, so that the method is economical in its storage requirements. The solution $\hat{x}_{k+1}$, on the other hand, is the best possible combination of the

directions $s_k$, but it is expensive to compute. It requires $O(k^2)$ inner products to form the normal equations (21.6). Moreover, we must maintain all the $s_j$ in memory or keep swapping them in and out of a backing store. It would therefore be desirable if we could find conditions on the $s_j$ that either improve the quality of the solution $x_{k+1}$ or make $\hat{x}_{k+1}$ easier to compute. In fact, there is a condition that does both.

11. We will say that the columns of $S_k$ are *A-conjugate* (or *conjugate* for short) if

$$S_k^{\mathrm{T}} A S_k = D_k,$$

where $D_k$ is diagonal. Since $S_k$ is of full rank, the diagonals of $D_k$ are nonzero — i.e., $D_k$ is nonsingular. Alternatively if we require $D_k$ to be nonsingular, then the rank of $S_k$ is $k$.

12. Conjugacy certainly simplifies the solution of the normal equations (21.6), since

$$\hat{a}_k = D_k^{-1} S_k^{\mathrm{T}} r_1$$

or

$$\hat{\alpha}_j = \frac{s_j^{\mathrm{T}} r_1}{s_j^{\mathrm{T}} A s_j}. \tag{21.8}$$

But more is true. Let

$$a_{j-1} = \begin{pmatrix} \alpha_1 \\ \alpha_2 \\ \vdots \\ \alpha_{j-1} \end{pmatrix}$$

be the vector of coefficients from the sequential minimization algorithm. Then

$$x_j = x_1 + S_{j-1} a_{j-1},$$

and

$$r_j = r_1 - A S_{j-1} a_{j-1}.$$

Hence by conjugacy

$$s_j^{\mathrm{T}} r_j = s_j^{\mathrm{T}} r_1 - s_j^{\mathrm{T}} A S_{j-1} a_{j-1} = s_j^{\mathrm{T}} r_1.$$

It follows from (21.7) and (21.8) that

$$\hat{\alpha}_j = \frac{s_j^{\mathrm{T}} r_1}{s_j^{\mathrm{T}} A s_j} = \alpha_j.$$

Thus the coefficients $\alpha_j$ generated by the sequential minimization are the same as the components of $\hat{a}$, and consequently $x_{k+1} = \hat{x}_{k+1}$. Hence if we work with conjugate directions we get the economies of sequential minimization and the optimality of global minimization.

## The method of conjugate gradients

13. We now need a supply of conjugate directions. One possibility is to note that conjugacy is equivalent to orthogonality in the inner product $(\cdot, \cdot)_A$ defined by

$$(x, y)_A = x^{\mathrm{T}} A y.$$

Consequently, we can apply the Gram–Schmidt algorithm to conjugate a set of independent vectors to get the $s_k$. Unfortunately, this requires about the same amount of work as the formation of the normal equations.

14. To get out of this bind, let's take a cue from the Lanczos algorithm, and consider what happens if we conjugate the Krylov sequence

$$s_1, As_1, A^2 s_1, \ldots.$$

In this case we can write

$$s_j = q_j(A) s_1,$$

where $q_j$ is a polynomial of degree $j - 1$. If we define an $A$-inner product over polynomials by

$$(p, q)_A = [p(A) s_1]^{\mathrm{T}} A [q(A) s_1],$$

then by conjugacy the polynomials $q_j$ are orthogonal in this inner product. Moreover, $(tp, q)_A = (p, tq)_A$. Hence, as in §20.10, these orthogonal polynomials satisfy a three-term recurrence relation, which makes their computation inexpensive.

15. Although sequential minimization along a set of directions obtained by conjugating a Krylov sequence is computationally inexpensive, it has an important drawback. If $s_1$ is an eigenvector of $A$, then $s_1$ and $As_1$ are linearly dependent, and the conjugation process terminates prematurely. We can avoid this problem, and at the same time simplify the formulas, by taking $s_1 = r_1$.

16. Let us suppose that we start with $s_1 = r_1$ and conjugate the Krylov sequence $s_1, As_1, A^2 s_1, \ldots$ to get the directions $s_1, s_2, \ldots$. The residuals $r_j$ are determined by the sequential minimization algorithm. As above let us denote the $k$th Krylov subspace by

$$\mathcal{K}_k = \operatorname{span}(r_1, Ar_1, \ldots, A^{k-1} r_1).$$

Then it is easy to see that

$$\mathcal{K}_k = \operatorname{span}(r_1, r_2, \ldots, r_k) = \operatorname{span}(s_1, s_2, \ldots, s_k).$$

Hence $r_k \in \mathcal{K}_k$. But $r_{k+1} \perp \mathcal{K}_k$, since $r_{k+1}$ is orthogonal to $s_1, \ldots, s_k$. Consequently, $r_{k+1}$ contains components along $A^k r_1$, and we can obtain $s_{k+1}$ by conjugating $r_{k+1}$ against $s_k, s_{k-1}, \ldots, s_1$. Thus we seek $s_{k+1}$ in the form

$$s_{k+1} = r_{k+1} - \beta_k s_k - \gamma_k s_{k-1} - \cdots.$$

The conjugacy condition $s_k^T A s_{k+1} = 0$ immediately gives the formula

$$\beta_k = \frac{s_k^T A r_{k+1}}{s_k^T A s_k}.$$  (21.9)

Similarly

$$\gamma_k = \frac{s_{k-1}^T A r_{k+1}}{s_{k-1}^T A s_{k-1}}.$$

But $A s_{k-1} \in \mathcal{K}_k \perp r_{k+1}$. Hence, $\gamma_k = 0$, and likewise all the way down. It follows that

$$s_{k+1} = r_{k+1} - \beta_k s_k,$$

where $\beta_k$ is given by (21.9).

17.   The results of the last few sections can be combined in the following *conjugate gradient algorithm*[28]

1.  $r_1 = s_1 = b - A x_1$
2.  **for** $k = 1, 2, \ldots$
3.     $\alpha_k = s_k^T r_k / s_k^T A s_k$
4.     $x_{k+1} = x_k + \alpha_k s_k$
5.     $r_{k+1} = r_k - \alpha_k A s_k$
6.     $\beta_k = s_k^T A r_{k+1} / s_k^T A s_k$
7.     $s_{k+1} = r_{k+1} - \beta_k s_k$
8.  **end for** $k$

Figure 21.1 exhibits the properties of the quantities generated by the conjugate gradient algorithm.

18. The formulas for $\alpha_k$ and $\beta_k$ are not the ones used in practice. Better formulas can be derived by exploiting the orthogonality and conjugacy relations.
   Multiplying the relation $s_k = r_k - \beta_{k-1} s_{k-1}$ by $r_k^T$ and using the fact that $r_k^T s_{k-1} = 0$, we find that $\|r_k\|_2^2 = s_k^T r_k$. Hence

$$\alpha_k = \frac{\|r_k\|_2^2}{s_k^T A s_k}.$$

   Multiplying the relation $r_{k+1} = r_k - \alpha_k A s_k$ by $r_{k+1}^T$ and using the orthogonality of the $r_j$, we get

$$\|r_{k+1}\|_2^2 = -\alpha_k r_{k+1}^T A s_k = -\frac{\|r_k\|_2^2}{s_k^T A s_k} r_{k+1}^T A s_k.$$

---

[28] The name is not quite right. The vectors $r_k$ are the gradients of the function $\varphi(x)$ at $x_k$. The vectors $s_k$ are obtained by conjugating the $r_k$. Thus the algorithm should properly be called the method of conju*gated* gradients.

1.  The vector $x_{k+1}$ minimizes $(x - x_*)^{\mathrm{T}} A(x - x_*)$ in the space $x_1 + \mathcal{K}_k$, where

$$\mathcal{K}_k = \mathrm{span}(r_1, Ar_1, \ldots, A^{k-1}r_1).$$

2.  The vectors $r_1, \ldots, r_k$ and $s_1, \ldots, s_k$ each span $\mathcal{K}_k$.
3.  The vectors $r_j$ are orthogonal and the vectors $s_j$ are conjugate.
4.  Equivalently

$$r_{k+1} \perp \mathcal{K}_k \quad \text{and} \quad s_{k+1} \perp_A \mathcal{K}_k,$$

where $\perp_A$ denotes conjugacy.

5.  The vector $r_k$ can be written in the form

$$r_k = p_k(A)r_1, \qquad (21.10)$$

where $p_k$ is the $k$th orthogonal polynomial in the inner product

$$(p, q) = [p(A)r_1]^{\mathrm{T}}[q(A)r_1].$$

6.  The vector $s_k$ can be written in the form

$$s_k = q_k(A)s_1,$$

where $q_k$ is the $k$th orthogonal polynomial in the inner product

$$(p, q)_A = [p(A)r_1]^{\mathrm{T}} A[q(A)r_1].$$

Figure 21.1. *Summary of the conjugate gradient algorithm.*

Dividing by $\|r_k\|_2^2$ we get

$$\beta_{k+1} = \frac{r_{k+1}^{\mathrm{T}} A s_k}{s_k^{\mathrm{T}} A s_k} = -\frac{\|r_{k+1}\|_2^2}{\|r_k\|_2^2}.$$

Thus the algorithm we actually use is

1.  $r_1 = s_1 = b - Ax_1$
2.  **for** $k = 1, 2, \ldots$
3.      $\alpha_k = \|r_k\|_2^2 / s_k^{\mathrm{T}} A s_k$
4.      $x_{k+1} = x_k + \alpha_k s_k$
5.      $r_{k+1} = r_k - \alpha_k A s_k$                    (21.11)
6.      $\beta_k = -\|r_{k+1}\|_2^2 / \|r_k\|_2^2$
7.      $s_{k+1} = r_{k+1} - \beta_k s_k$
8.  **end for** $k$

The reason these formulas are superior is that the norms are always accurately computed, whereas cancellation can occur in the computation of $s_k^{\mathrm{T}} r_k$ and $s_k^{\mathrm{T}} A r_{k+1}$.

## Termination

19. We have already noted (§21.15) that if $s_1$ is an eigenvector of $A$ then the conjugation process never gets off the ground. However, if $s_1 = r_1$, we have for some $\lambda > 0$ (remember that $A$ is positive definite),

$$e_1 = -A r_1 = -\lambda r_1 = -\lambda s_1.$$

Thus the first step is along the direction of the error, which is immediately reduced to zero.

20. An analogous termination of the procedure occurs when $\mathcal{K}_k$ is an invariant subspace of $A$. Specifically, let $m$ be the smallest integer such that $\mathcal{K}_m = \mathcal{K}_{m+1}$, i.e., such that $A^m r_1$ is a linear combination of $r_1, \ldots, A^{m-1} r_1$. Then

$$A\mathcal{K}_m \subset \mathcal{K}_{m+1} = \mathcal{K}_m,$$

so that $\mathcal{K}_m$ is an invariant subspace of $A$. Now $r_{m+1} \perp \mathcal{K}_m$. But $r_{m+1} \in \mathcal{K}_{m+1} = \mathcal{K}_m$. Thus $r_{m+1}$ is in and is orthogonal to $\mathcal{K}_m$ and hence must be zero. We have established the following result.

> Let $m$ be the smallest integer such that
> $$\mathcal{K}_m = \mathcal{K}_{m+1}.$$
> Then $x_{m+1} = x_*$.

Since we always have $\mathcal{K}_n = \mathcal{K}_{n+1}$, we have shown that:

> The conjugate gradient method terminates with a solution in at most $n$ iterations.

# Krylov Sequence Methods

Operation Counts and Storage Requirements
Conjugate Gradients as an Iterative Method
Convergence in the $A$-Norm
Monotone Convergence in the 2-Norm

## Operation counts and storage requirements

1. The key operation in the method of conjugate gradients is the formation of the product $As_k$. If $A$ is dense, this operation requires about $n^2$ additions and multiplications. Hence if the process does not terminate before the $n$th step, the algorithm requires $n^3$ additions and multiplications, the other manipulations in the algorithm being of a lower order. This should be compared with Gaussian elimination, which requires $\frac{1}{3}n^3$.

2. If $A$ is sparse or structured, the number of operations required to form $Ax$ will depend on the matrix itself. If it is small enough, the computations in the algorithm itself become important. Referring to (21.11), we see that the calculation of $\alpha_k$ and $\beta_k$ requires $n$ additions and multiplication for each (the value of $\|r_k\|_2$ was computed in the preceding iteration). Likewise the updating of $x_k$, $r_k$, and $s_k$ requires $n^2$ additions and multiplications for each. Thus, each step of the algorithm requires $5n^2$ additions and multiplications, aside from the matrix-vector product.

3. The algorithm requires that we store the components of $x_k$, $r_k$, and $s_k$, which may be overwritten by their successors. In addition we must store $As_k$. Thus the overall storage requirement is $4n$ floating-point words.

## Conjugate gradients as an iterative method

4. We have just seen that for dense matrices the method of conjugate gradients cannot compete with Gaussian elimination. Because the individual steps of the algorithm are cheap, it is a natural for large sparse matrices. But it would be impossibly expensive to push the method to termination. For this reason conjugate gradients is used exclusively as an iterative method that produces increasingly accurate approximations to the solution. We will now investigate the rate of convergence.

## Convergence in the A-norm

5. Since the conjugate gradient method minimizes the error of the current iterate in the $A$-norm defined by

$$\|x\|_A^2 = x^{\mathrm{T}} Ax,$$

191

it is natural to try to bound it. Recall that [see (21.10)]

$$r_{k+1} = p_k(A)r_1,$$

where $p_k$ is a polynomial of degree $k$. Since

$$r_k = r_1 + \text{terms in } A, A^2, \ldots,$$

the constant term of $p_k$ must be one; that is, $p_k(0) = 1$. If we multiply by $A^{-1}$, we find that

$$e_{k+1} = p_k(A)e_1$$

satisfies the same relation [see (21.2)].

Now suppose that we take arbitrary steps along the directions $s_k$. Then the error at the $k$th step may be represented in the form

$$\tilde{e}_{k+1} = p(A)e_1,$$

where $p$ is a polynomial of degree not greater than $k$ satisfying

$$p(0) = 1. \tag{22.1}$$

Since $e_{k+1}^{\mathrm{T}} A e_{k+1} = \|e_{k+1}\|_A^2$ is as small as possible over $\mathcal{K}_k$,

$$\|e_{k+1}\|_A \le \|p(A)e_1\|_A$$

for any polynomial satisfying $p(0) = 1$. Thus we may try to choose a polynomial $p$ satisfying (22.1), in such a way that the above inequality gives a good bound on $\|e_{k+1}\|_A$.

6. We begin by relating the error to the eigenvalues of $A$. Since $A$ is positive definite, it has a positive definite square root $A^{\frac{1}{2}}$. Now for any vector $x$,

$$\|x\|_A = \|A^{\frac{1}{2}}x\|_2.$$

Hence

$$\|p(A)e_1\|_A = \|p(A)A^{\frac{1}{2}}e_1\|_2 \le \|p(A)\|_2 \|A^{\frac{1}{2}}e_1\|_2 = \|p(A)\|_2 \|e_1\|_A.$$

But the 2-norm of any symmetric matrix is the magnitude of its largest eigenvalue. Hence if we denote the set of eigenvalues of $A$ by $\lambda(A)$,

$$\|p(A)e_1\|_A = \max_{\lambda \in \lambda(A)} |p(\lambda)| \, \|e_0\|_A.$$

7. Ideally we would like to choose $p$ to minimize $\max_{\lambda \in \lambda(A)} |p(\lambda)|$. If we know something about the distribution of the eigenvalues of $A$ we can try to tailor $p$ so that the maximum is small. Here we will suppose that we know only the

smallest and largest eigenvalues of $A$ — call them $a$ and $b$. We then wish to determine $p$ so that

$$\max_{t\in[a,b]} |p(t)| = \min \tag{22.2}$$

subject to the constraint that $p(0) = 1$.

8. All the techniques we need to solve this problem are at hand. Suppose that $p$ is a polynomial of degree $k$ with $p(0) = 1$ that equi-alternates $k + 1$ times in $[a, b]$. Then by a modification of the proof in §3.6 we can show that this polynomial satisfies (22.2). But such a polynomial is given by

$$p(t) = \frac{c_k \left( \frac{t-a}{b-a} + \frac{t-b}{b-a} \right)}{c_k \left( -\frac{b+a}{b-a} \right)}, \tag{22.3}$$

where $c_k$ is the $k$th Chebyshev polynomial. Since the numerator in (22.3) equi-alternates between $\pm 1$, the problem of computing a bound reduces to evaluating the denominator of (22.3).

9. Let $\kappa = b/a$. Then

$$\left| c_k \left( -\frac{b+a}{b-a} \right) \right| = c_k \left( \frac{\kappa + 1}{\kappa - 1} \right).$$

Now it is easily verified that

$$\cosh^{-1} \left( \frac{\kappa + 1}{\kappa - 1} \right) = -\ln \frac{\sqrt{\kappa} - 1}{\sqrt{\kappa} + 1} \equiv -\ln \sigma.$$

Since $c_k(t) = \cosh(k \cosh^{-1} t)$, we have

$$\frac{1}{c_k \left( \frac{b+a}{b-a} \right)} = \frac{2\sigma^k}{1 + \sigma^{2k}} \leq 2\sigma^k.$$

10. If we recall that $\kappa = b/a = \|A\|_2 \|A^{-1}\|_2$ is the condition number of $A$, we may summarize all this as follows.

> Let $\kappa = \|A\|_2 \|A^{-1}\|_2$ be the condition number of $A$. Then the errors $e_k$ in the conjugate gradient method satisfy
>
> $$\|e_k\|_A \leq 2 \left( \frac{\sqrt{\kappa} - 1}{\sqrt{\kappa} + 1} \right)^k \|e_1\|_A. \tag{22.4}$$

## Monotone convergence in the 2-norm

11. Since all norms are equivalent, in the sense that convergence in one norm implies convergence in any other, the bound (22.4) formally establishes the convergence of the conjugate gradient method in the 2-norm. Unfortunately, it is possible for the $A$-norm of a vector to be small, while its 2-norm is large. For example, suppose

$$A = \begin{pmatrix} 1 & 0 \\ 0 & \eta \end{pmatrix},$$

where $\eta$ is small. Then $\|x\|_A = \xi_1^2 + \eta\xi_2^2$. Thus if $\|x\|_A = 1$, it is possible for $|\xi_2|$ to be as large as $1/\sqrt{\eta}$. It is even possible for the $A$-norm of a vector to decrease while its 2-norm increases. Given such phenomena, the optimality of the conjugate gradient method seems less compelling. What good does it do us to minimize the $A$-norm of the error if its 2-norm is increasing? Fortunately, the 2-norm also decreases, a result which we shall now establish.

12. Let's begin by establishing a relation between $\|e_k\|_2$ and $\|e_{k+1}\|_2$. From the equation

$$x_{k+1} = x_k + \alpha_k s_k,$$

we have $e_{k+1} = e_k + \alpha_k s_k$ or

$$e_k = e_{k+1} - \alpha_k s_k.$$

Hence

$$\|e_k\|_2^2 = (e_{k+1} - \alpha_k s_k)^{\mathrm{T}}(e_{k+1} - \alpha_k s_k) = \|e_{k+1}\|_2^2 - 2\alpha_k s_k^{\mathrm{T}} e_{k+1} + \alpha^2\|s_k\|_2^2,$$

or

$$\|e_{k+1}\|_2^2 = \|e_k\|_2^2 + 2\alpha_k s_k^{\mathrm{T}} e_k - \alpha^2\|s_k\|_2^2.$$

Now the term $-\alpha^2\|s_k\|_2^2$ reduces the error. Since $\alpha_k = \|r_k\|_2^2/s_k^{\mathrm{T}} A s_k > 0$, the term $2\alpha_k s_k^{\mathrm{T}} e_{k+1}$ will reduce the error provided $s_k^{\mathrm{T}} e_{k+1}$ is not positive.

13. Actually we will show a great deal more than the fact that $s_k^{\mathrm{T}} e_{k+1} \leq 0$. We will show that the inner product of any error with any step is nonpositive. The approach is to use matrix representations of the conjugate gradient recursions.

14. Let $m$ be the iteration at which the conjugate gradient algorithm terminates, so that $r_{m+1} = 0$. For purposes of illustration we will take $m = 4$.

Write the relation $s_{k+1} = r_{k+1} - \beta_k s_k$ in the form

$$r_{k+1} = s_{k+1} + \beta_k s_k.$$

Recalling that $r_1 = s_1$, we can write these relations in the form

$$(r_1 \ r_2 \ r_3 \ r_4) = (s_1 \ s_2 \ s_3 \ s_4) \begin{pmatrix} 1 & \beta_1 & 0 & 0 \\ 0 & 1 & \beta_2 & 0 \\ 0 & 0 & 1 & \beta_3 \\ 0 & 0 & 0 & 1 \end{pmatrix}.$$

We abbreviate this by writing

$$R = SU, \qquad (22.5)$$

defining $R$, $S$, and $U$. Note for future reference that since

$$\beta_k = -\frac{\|r_{k+1}\|_2}{\|r_k\|_2} < 0,$$

the matrix $U$ is upper triangular with positive diagonal and nonpositive off-diagonal elements.

15. Now write the relation $r_{k+1} = r_k - \alpha_k A s_k$ in the form

$$\alpha_k A s_k = r_k - r_{k+1}.$$

In terms of matrices, these relations become

$$A(s_1 \; s_2 \; s_3 \; s_4) \begin{pmatrix} \alpha_1 & 0 & 0 & 0 \\ 0 & \alpha_2 & 0 & 0 \\ 0 & 0 & \alpha_3 & 0 \\ 0 & 0 & 0 & \alpha_4 \end{pmatrix} = (r_1 \; r_2 \; r_3 \; r_4) \begin{pmatrix} 1 & 0 & 0 & 0 \\ -1 & 1 & 0 & 0 \\ 0 & -1 & 1 & 0 \\ 0 & 0 & -1 & 1 \end{pmatrix}.$$

Here we have used the fact that $r_5 = 0$. We abbreviate this equation as

$$ASD = RL. \qquad (22.6)$$

Since $\alpha_k = \|r_k\|_2^2 / s_k^{\mathrm{T}} A s_k$, the matrix $D$ has positive diagonal elements. Moreover, $L$ is a lower triangular matrix with positive diagonal and nonpositive off-diagonal elements.

16. The reason for stressing the nature of the triangular matrices in the above relations is the following elementary but extremely useful result.

> If $T$ is a triangular matrix of order $n$ with positive diagonal and nonpositive off-diagonal elements, then the elements of $T^{-1}$ are nonnegative.

To establish this fact write $T = \Delta - C$, where $\Delta$ is the diagonal of $T$. Then it is easy to verify that $(\Delta^{-1}C)^n = 0$ and hence that

$$T^{-1} = [I + \Delta^{-1}C + \cdots + (\Delta^{-1}C)^{n-1}]\Delta^{-1}.$$

But all the terms in this representation are nonnegative.

17. Returning to our relations, it follows from (22.5) that

$$S^{\mathrm{T}}S = U^{-\mathrm{T}}R^{\mathrm{T}}RU^{-1}.$$

But the columns of $R$ are orthogonal, and hence $R^{\mathrm{T}}R$ is a diagonal matrix with positive diagonal entries. Moreover by the result of §22.16, the matrix $U^{-1}$ is nonnegative. Hence $S^{\mathrm{T}}S$ is nonnegative.

Now let

$$E = (e_1 \; e_2 \; e_3 \; e_4)$$

be the matrix of errors. Then $E = -A^{-1}R$. Hence from (22.6),

$$E = -SDL^{-1}.$$

On multiplying by $S^{\mathrm{T}}$, we have

$$S^{\mathrm{T}}E = -(S^{\mathrm{T}}S)DL^{-1}.$$

The matrices $S^{\mathrm{T}}S$, $D$, and $L$ are all nonnegative. Hence the elements of $S^{\mathrm{T}}E$ are nonpositive. We may summarize as follows.

---

In the conjugate gradient algorithm, the inner products $s_i^{\mathrm{T}}s_j$ are nonnegative and the inner products $s_i^{\mathrm{T}}e_j$ are nonpositive. The quantity $2\alpha_k s_k^{\mathrm{T}}e_k$ is nonpositive, and

$$\|e_{k+1}\|_2^2 = \|e_k\|_2^2 - \alpha^2\|s_k\|_2^2 + 2\alpha_k s_k^{\mathrm{T}}e_k.$$

Consequently, the conjugate gradient method converges monotonically in the 2-norm.

---

# Krylov Sequence Methods

Preconditioned Conjugate Gradients
Preconditioners
Incomplete LU Preconditioners

## Preconditioned conjugate gradients

1. We have seen that the $A$-norm of the error in the conjugate gradient iterates converges to zero at least as fast as

$$\left(\frac{\sqrt{\kappa}-1}{\sqrt{\kappa}+1}\right)^{k},$$

where $\kappa$ is the condition number of $A$. If $\kappa = 1$ the method converges in one iteration. On the other hand, if the condition number is large, the convergence can be extremely slow. The idea behind preconditioning is to replace the original problem with a better conditioned one — i.e., one in which the matrix of the system is nearer the perfectly conditioned identity matrix.

2. Since $A$ itself is fixed, we have to fiddle with the system to move its matrix toward the identity. Let $M$ be a positive definite matrix (called a *preconditioner*) that in some sense approximates $A$. (We will worry about how to find such approximations later.) Then it is natural to consider solving the system $(M^{-1}A)x = M^{-1}b$. Unfortunately, this system is no longer symmetric, so that we cannot use conjugate gradients. However, because $M$ is positive definite, it has a positive definite square root, and we can try to solve the system

$$(M^{-\frac{1}{2}}AM^{-\frac{1}{2}})(M^{\frac{1}{2}}x) = (M^{-\frac{1}{2}}b), \tag{23.1}$$

whose matrix is positive definite.

3. It is not practical to plug (23.1) into the conjugate gradient algorithm, since $M^{\frac{1}{2}}$ will not generally be available. Besides the result will not be $x_*$, which is what we want, but $M^{\frac{1}{2}}x_*$, which we don't particularly care about. Nonetheless, let's write down the conjugate gradient method for this system, and see if we can transform it into something more useful.

1.  $\hat{r}_1 = (M^{-\frac{1}{2}}b) - (M^{-\frac{1}{2}}AM^{-\frac{1}{2}})(M^{\frac{1}{2}}x_1)$
2.  $\cdots$
3.  $\hat{s}_1 = \hat{r}_1$
4.  **for** $k = 1, 2, \ldots$
5.  $\qquad \alpha_k = \|\hat{r}_k\|_2^2 / \hat{s}_k^{\mathrm{T}}(M^{-\frac{1}{2}}AM^{-\frac{1}{2}})\hat{s}_k$
6.  $\qquad (M^{\frac{1}{2}}x_{k+1}) = (M^{\frac{1}{2}}x_k) + \alpha_k\hat{s}_k$
7.  $\qquad \hat{r}_{k+1} = \hat{r}_k - \alpha_k(M^{-\frac{1}{2}}AM^{-\frac{1}{2}})\hat{s}_k$
8.  $\qquad \cdots$
9.  $\qquad \beta_k = -\|\hat{r}_{k+1}\|_2^2 / \|\hat{r}_k\|_2^2$
10. $\qquad \hat{s}_{k+1} = \hat{r}_{k+1} - \hat{\beta}_k\hat{s}_k$
11. **end for** $k$

The quantities with hats over them will change in the final algorithm. Note the two blank lines, which will be filled in later.

The logical place to begin transforming this algorithm is with the updating of $x_k$. If we multiply statement 6 by the matrix $M^{-\frac{1}{2}}$ and set

$$s_k = M^{-\frac{1}{2}}\hat{s}_k,$$

the result is

6.  $\qquad x_{k+1} = x_k + \alpha_k s_k$

Now for the residuals. If we multiply statement 1 by $M^{\frac{1}{2}}$ and set

$$r_k = M^{\frac{1}{2}}\hat{r}_k,$$

we get

1.  $\qquad r_1 = b - Ax_1$

Similarly statement 7 becomes

7.  $\qquad r_{k+1} = r_k - \alpha_k As_k$

So far we have merely reproduced statements in the original conjugate gradient algorithm. Something different happens when we try to update $s_k$. Consider first $s_1$. If we multiply statement 3 by $M^{\frac{1}{2}}$ to transform $\hat{r}_1$ into $r_1$, the right-hand side of the assignment becomes $M^{\frac{1}{2}}\hat{s}_1 = Ms_1$. Thus $s_1 = M^{-1}r_1$. We can get around this problem by defining

$$z_k = M^{-1}r_k$$

and replacing statements 2 and 3 by

    2.    Solve the system $Mz_1 = r_1$

    3.    $s_1 = z_1$

Similarly statements 8 and 10 become

    8.        Solve the system $Mz_{k+1} = r_{k+1}$

and

    10.      $s_{k+1} = z_{k+1} + \beta_k s_k$

Finally, we need an expression for $\|\hat{r}_k\|_2^2$. Since $\hat{r}_k = M^{-\frac{1}{2}} r_k$, we have

$$\|\hat{r}_k\|_2^2 = r_1^{\mathrm{T}} M^{-1} r_1 = r_1^{\mathrm{T}} z_1.$$

Consequently statements 5 and 9 assume the forms

    5.      $\alpha_k = r_k^{\mathrm{T}} z_k / s_k^{\mathrm{T}} A s_k$

and

    9.      $\beta_k = -r_{k+1}^{\mathrm{T}} z_{k+1} / r_k^{\mathrm{T}} z_k$

4. If we collect the above changes into a single algorithm, we get the method of *preconditioned conjugate gradients*.

    1.    $r_1 = b - Ax_1$
    2.    Solve the system $Mz_1 = r_1$
    3.    $s_1 = z_1$
    4.    **for** $k = 1, 2, \ldots$
    5.        $\alpha_k = r_k^{\mathrm{T}} z_k / s_k^{\mathrm{T}} A s_k$
    6.        $x_{k+1} = x_k + \alpha_k s_k$
    7.        $r_{k+1} = r_k - \alpha_k A x_k$
    8.        Solve the system $Mz_{k+1} = r_{k+1}$
    9.        $\beta_k = -r_{k+1}^{\mathrm{T}} z_{k+1} / r_k^{\mathrm{T}} z_k$
    10.      $s_{k+1} = z_{k+1} + \beta_k s_k$
    11.  **end for** $k$

Except for the solution of the system $Mz_k = r_k$ this algorithm is no more costly than ordinary conjugate gradients.

## Preconditioners

5. There are two desiderata in choosing a preconditioner $M$.

    1.    Systems of the form $Mz = r$ must be easy to solve.

    2.    The matrix $M$ must approximate the matrix $A$.

These two desiderata generally work against each other. For example choosing $M = I$ makes the system easy to solve but gives a poor approximation to $A$ (unless of course $A \cong I$, in which case we do not need to precondition). On the other hand, the choice $M = A$ causes the conjugate gradient algorithm to converge in one step but at the expense of solving the original problem with a different right-hand side. We must therefore compromise between these two extremes.

6. A choice near the extreme of choosing $M = I$ is to chose $M = D$, where $D$ is the diagonal of $A$. The system $Mz = r$ is trivial to solve. Moreover, if the elements of $D$ vary over many orders of magnitude, this *diagonal preconditioning* may speed things up dramatically.

It might be objected that since $D$ is diagonal it is easy to form the matrix $D^{-\frac{1}{2}}AD^{-\frac{1}{2}}$. Why then go through the trouble of using the preconditioned version of an algorithm? An answer is that in some circumstances we may not be able to manipulate the matrix. We have already noted that the matrix $A$ enters into the algorithm only in the formation of the product $Ax$. If this computation is done in a black-box subprogram, it may be easier to use preconditioned conjugate gradients than to modify the subprogram.

7. Another possibility is to use larger pieces of $A$. For example, a matrix which arises in the solution of Poisson's problem has the form

$$
\begin{pmatrix}
T & -I & & & & \\
-I & T & -I & & & \\
& -I & T & -I & & \\
& & \ddots & \ddots & \ddots & \\
& & & -I & T & -I \\
& & & & -I & T
\end{pmatrix},
\tag{23.2}
$$

where

$$
T =
\begin{pmatrix}
4 & -1 & & & & \\
-1 & 4 & -1 & & & \\
& -1 & 4 & -1 & & \\
& & \ddots & \ddots & \ddots & \\
& & & -1 & 4 & -1 \\
& & & & -1 & 4
\end{pmatrix}.
$$

In this case a natural preconditioner is

$$
M = \mathrm{diag}(T, T, T, \dots, T, T).
$$

This preconditioner has more of the flavor of $A$ than the diagonal of $A$. And since the matrix $T$ is tridiagonal, it is cheap to solve systems involving $M$.

## Incomplete LU preconditioners

8. Except for diagonal matrices, the solution of the system $Mz = r$ requires that we have a suitable decomposition of $M$. In many instances this will be an $LU$ decomposition — the factorization $M = LU$ into the product of a unit lower triangular matrix and an upper triangular matrix produced by Gaussian elimination. The idea of an incomplete LU preconditioner is to perform an abbreviated form of Gaussian elimination on $A$ and to declare the product of the resulting factors to be $M$. Since $M$ is by construction already factored, systems involving $M$ will be easy to solve.

9. To get an idea of how the process works, consider the matrix (23.2). Since we will be concerned with only the structure of this matrix, we will use a Wilkinson diagram to depict an initial segment:

$$
\begin{pmatrix}
X & X & & & X & & & \\
X & X & X & & & X & & \\
& X & X & X & & & X & \\
& & X & X & & & & X \\
X & & & & X & X & & \\
& X & & & X & X & X & \\
& & X & & & X & X & X \\
& & & X & & & X & X \\
\end{pmatrix}.
$$

Now suppose we perform Gaussian elimination on this matrix. The elimination below the diagonal will cause elements above the diagonal to fill in, giving the matrix

$$
\begin{pmatrix}
X & X & & & X & & & \\
0 & X & X & & X & X & & \\
& 0 & X & X & X & X & X & \\
& & 0 & X & X & X & X & X \\
0 & & & & X & X & & \\
& 0 & & & X & X & X & \\
& & 0 & & X & X & X & X \\
& & & 0 & X & X & X & X \\
\end{pmatrix}.
$$

If we were concerned with factoring $A$, this fill-in would be fatal.[29] However, suppose that we propagate the effects of eliminating an element selectively across the row of the matrix. For example, if an element in the row in question is nonzero, we might treat it normally; if it is zero, we might do nothing, leaving it zero. The resulting matrices $L$ and $U$ will have exactly the same pattern of nonzero elements as $A$. Moreover, their product $LU$ will reproduce the nonzero elements of $A$ exactly; however, it will not reproduce the zero elements of $A$. This is an example of an *incomplete LU factorization*.

---

[29]There are better ways of ordering the elements of this matrix for elimination — nested dissection for one.

10. It is now time to get precise. The sparsity pattern of $L$ and $U$ does not necessarily have to be the same as $A$ (as it was above). We will therefore introduce a *sparsity set* $\mathcal{Z}$ to control the pattern of zeros. Specifically, let $\mathcal{Z}$ be a set of ordered pairs of integers from $\{1, 2, \ldots, n\}$ containing no pairs of the form $(i, i)$. An incomplete LU factorization of $A$ is a decomposition of the form

$$A = LU + R, \qquad (23.3)$$

where $L$ is unit lower triangular (i.e., lower triangular with ones on its diagonal), $U$ is upper triangular, and $L$, $U$, and $R$ have the following properties.

1. If $(i, j) \in \mathcal{Z}$ with $i > j$, then $\ell_{ij} = 0$.
2. If $(i, j) \in \mathcal{Z}$ with $i < j$, then $u_{ij} = 0$.
3. If $(i, j) \notin \mathcal{Z}$, then $r_{ij} = 0$.

In other words, the elements of $L$ and $U$ are zero on the sparsity set $\mathcal{Z}$, and off the sparsity set the decomposition reproduces $A$.

11. It is instructive to consider two extreme cases. If the sparsity $\mathcal{Z}$ set is empty, we get the LU decomposition of $A$ — i.e., we are using $A$ as a preconditioner. On the other hand, if $\mathcal{Z}$ is everything except diagonal pairs of the form $(i, i)$, then we are effectively using the diagonal of $A$ as a preconditioner.

12. To establish the existence and uniqueness of the incomplete factorization, we will first give an algorithm for computing it, which will establish the uniqueness. We will then give conditions that insure existence and hence that the algorithm goes to completion.

13. The algorithm generates $L$ and $U$ rowwise. Suppose we have computed the first $k - 1$ rows of $L$ and $U$, and we wish to compute the $k$th row. Write the first $k$ rows of (23.3) in the form (northwest indexing)

$$\begin{pmatrix} A_{11} & A_{1k} \\ a_{k1}^{\mathrm{T}} & a_{kk}^{\mathrm{T}} \end{pmatrix} = \begin{pmatrix} L_{11} & 0 \\ \ell_{1k}^{\mathrm{T}} & 1 \end{pmatrix} \begin{pmatrix} U_{11} & U_{1k} \\ 0 & u_{kk}^{\mathrm{T}} \end{pmatrix} + \begin{pmatrix} R_{11} & R_{1k} \\ r_{k1}^{\mathrm{T}} & r_{kk}^{\mathrm{T}} \end{pmatrix}. \qquad (23.4)$$

We need to compute $\ell^{\mathrm{T}}$ and $u^{\mathrm{T}}$. Multiplying out (23.4), we find that

$$\ell_{k1}^{\mathrm{T}} U_{11} + r_{k1}^{\mathrm{T}} = a_{k1}^{\mathrm{T}} \qquad (23.5)$$

and

$$u_{kk}^{\mathrm{T}} + r_{kk}^{\mathrm{T}} = a_{kk}^{\mathrm{T}} - \ell_{k1}^{\mathrm{T}} U_{1k}. \qquad (23.6)$$

We are going to solve these two systems in the order

$$\ell_{k1}, \ldots, \ell_{k,k-1}, v_{k,k}, \ldots, v_{k,n}.$$

```
1.    incompleteLU(A, Z)
2.       for k = 1 to n
3.          for j = 1 to k−1
4.             if ((k, j) ∈ Z)
5.                L[k, j] = 0
6.             else
7.                L[k, j] = (A[k, j]−L[k, 1:j−1]*U[1:j−1, j])/U[j, j]
8.             end if
9.          end for j
10.         for j = k to n
11.            if ((k, j) ∈ Z)
12.               U[k, j] = 0
13.            else
14.               U[k, j] = A[k, j]−L[k, 1:k−1]*U[1:k−1, j]
15.            end if
16.         end for j
17.      end for k
18.   end incompleteLU
```

Figure 23.1. *Incomplete LU decomposition.*

Suppose that we have computed $\ell_{k1}, \ldots, \ell_{k,j-1}$. If $(k, j)$ is in the sparsity set, we simply set $\ell_{kj} = 0$. If $(k, j)$ is not in the sparsity set, then $r_{kj} = 0$, and (23.5) gives

$$\alpha_{kj} = \sum_{i=1}^{k-1} \ell_{ki} v_{ij} + \ell_{kj} v_{jj},$$

from which we get

$$\ell_{kj} = \frac{\alpha_{kj} - \sum_{i=1}^{k-1} \ell_{ki} v_{ij}}{v_{jj}}. \tag{23.7}$$

The key observation here is that it does not matter how the values of the preceding $\ell$'s and $v$'s were determined. If $\ell_{kj}$ is defined by (23.7), then when we compute $L$ its $(k, j)$-element will be $\alpha_{kj}$. Thus we can set $\ell$'s and $v$'s to zero on the sparsity set without interfering with the values of $LU$ off the sparsity set. A similar analysis applies to the determination of $v_{kk}, \ldots, v_{kn}$ from (23.6).

14. Figure 23.1 contains an implementation of the algorithm sketched above. If $Z$ is the empty set, the algorithm produces an LU factorization of $A$, and hence it is a variant of classical Gaussian elimination. When $Z$ is nonempty, the algorithm produces matrices $L$ and $U$ with the right sparsity pattern. Moreover, it must reproduce $A$ off the sparsity set, since that is exactly what statements 7 and 14 say.

The algorithm can be carried to completion provided the quantities $U[j, j]$ are all nonzero, in which case the decomposition is unique. Whether or not the $U[j, j]$ are nonzero will depend on the matrix in question. We are now going to investigate classes of matrices for which the algorithm always works.

# Krylov Sequence Methods

Diagonally Dominant Matrices
Return to Incomplete Factorization

## Diagonally dominant matrices

1. The matrix (23.2) associated with Poisson's problem is typical of many matrices in positive definite systems in that most of the matrix is concentrated on the diagonal. Such matrices are said to be diagonally dominant. It is also nonsingular. The goal of this lecture is to show that if $A$ is a nonsingular, diagonally dominant matrix then $A$ has an incomplete LU factorization for any sparsity set.

2. We begin with a definition. A matrix $A$ of order $n$ is *diagonally dominant* if

$$|a_{ii}| \geq \sum_{\substack{j=1 \\ j \neq i}}^{n} |a_{ij}|, \qquad i = 1, \ldots, n.$$

It is *strictly diagonally dominant* if strict inequality holds for all $j$.

3. The definition of diagonal dominance is a rowwise definition. We could also give a columnwise definition. For our purposes it will make no difference whether a matrix is diagonally dominant by rows or by columns.

4. Although the definition of diagonal dominance is simple, it has far-reaching consequences. We begin with the following observation on singular diagonally dominant matrices.

---

Let $A$ be diagonally dominant, and suppose that

$$Ax = 0. \tag{24.1}$$

If

$$|x_i| = \max_j\{|x_j|\} > 0, \tag{24.2}$$

then

$$a_{ij} \neq 0 \implies |x_j| = |x_i|. \tag{24.3}$$

Moreover,

$$|a_{ii}| = \sum_{\substack{j=1 \\ j \neq i}}^{n} |a_{ij}|. \tag{24.4}$$

---

# 206

# Afternotes Goes to Graduate School

To establish this result, note that from (24.1)

$$0 = a_{ii} + \sum_{j \neq i} a_{ij} \frac{x_j}{x_i}.$$

Hence

$$0 \geq |a_{ii}| - \sum_{j \neq i} |a_{ij}| \frac{|x_j|}{|x_i|}$$
$$\geq |a_{ii}| - \sum_{j \neq i} |a_{ij}|,$$

the last inequality following from (24.2). Moreover, the second inequality is strict inequality if and only if for some $j$ we have $|a_{ij}| > 0$ and $|x_j|/|x_i| < 1$. But strict inequality would contradict diagonal dominance. Hence (24.3) and (24.4).

5. The above result implies that if $A$ is strictly diagonal dominant then we cannot have $Ax = 0$ for a nonzero $x$. For if so, the inequality cannot be strict in any row corresponding to a maximal component of $x$. Thus we have shown that:

---
A strictly diagonally dominant matrix is nonsingular.

---

6. Although strictly diagonally dominant matrices must be nonsingular, diagonal dominance alone does not imply either nonsingularity or singularity. For example, the matrix

$$\begin{pmatrix} 2 & -1 & 0 \\ -1 & 2 & -1 \\ 0 & -1 & 2 \end{pmatrix}$$

is nonsingular. (Check it out by your favorite method.) On the other hand, the matrix

$$\begin{pmatrix} 1 & 1 & 0 \\ 1 & 1 & 0 \\ 1 & 2 & 4 \end{pmatrix}$$

is singular. Note that its null vector is $(1\ -1\ 1/4)$. Since the third component is not maximal whereas the first two are, by the result of §24.4 the two elements of $A$ in the northeast must be zero — as indeed they are.

7. If we can verify that a diagonally dominant matrix is nonsingular, we can generate a lot of other nonsingular diagonally matrices.

---
Let $A$ be diagonally dominant and nonsingular. Let $\hat{A}$ be obtained from $A$ by decreasing the magnitudes of a set (perhaps empty) of off-diagonal elements and increasing the magnitudes of a set (also perhaps empty) of diagonal elements. Then $\hat{A}$ is nonsingular.

---

To establish this result, suppose that $\hat{A}x = 0$, where $x \neq 0$. Now if we interchange two rows and the same two columns of $\hat{A}$, the diagonals are interchanged along with the elements in their respective rows (though the latter appear in a different order). Moreover, in the equation $\hat{A}x = 0$, the corresponding components of $x$ are interchanged. By a sequence of such interchanges, we may bring the maximal components of $x$ to the beginning.

Assuming such interchanges have been made, partition $Ax = 0$ in the form

$$\begin{pmatrix} \hat{A}_{11} & \hat{A}_{12} \\ \hat{A}_{21} & \hat{A}_{22} \end{pmatrix} \begin{pmatrix} x_1 \\ x_2 \end{pmatrix} = \begin{pmatrix} 0 \\ 0 \end{pmatrix},$$

where $x_1$ contains all the maximal components of $x$. By the result of §24.4, the matrix $\hat{A}_{12}$ must be zero, so that $\hat{A}$ has the form

$$\begin{pmatrix} \hat{A}_{11} & 0 \\ \hat{A}_{21} & \hat{A}_{22} \end{pmatrix}, \qquad (24.5)$$

where $A_{11}$ is singular and diagonally dominant with equality in all rows.

Now in passing back to $A$, we must decrease the magnitude of some diagonals and increase the magnitude of some off-diagonals. Some of these modifications must take place in the first row of the partition (24.5), for otherwise $A$ will be singular. But any such modification will destroy the diagonal dominance of $A$.

8. It is worth pointing out that this result is about changing magnitudes, independently of signs. Thus we may change the sign of an off-diagonal element as long as the magnitude decreases. But in general we cannot do without a strict decrease, as the matrices

$$\begin{pmatrix} 1 & 1 \\ 1 & -1 \end{pmatrix} \quad \text{and} \quad \begin{pmatrix} 1 & 1 \\ 1 & 1 \end{pmatrix}$$

show.

9. An important consequence of the result of §24.7 is that:

> Any principal submatrix of a nonsingular diagonally dominant matrix is a nonsingular diagonally dominant matrix.

To see this assume without loss of generality that the principal submatrix is a leading submatrix, say $A_{11}$ in the partition

$$A = \begin{pmatrix} A_{11} & A_{12} \\ A_{21} & A_{22} \end{pmatrix}.$$

Then $A_{11}$ is clearly diagonally dominant. Moreover, by the result of §24.7, the matrix

$$\begin{pmatrix} A_{11} & 0 \\ 0 & A_{22} \end{pmatrix}$$

is nonsingular, which can only be true if $A_{11}$ is nonsingular.

10. We are now going to investigate the relation of Gaussian elimination and diagonal dominance. To fix our notation, suppose we want to perform one step of Gaussian elimination on the matrix

$$A = \begin{pmatrix} \alpha_{11} & a_{12}^{\mathrm{T}} \\ a_{21} & A_{22} \end{pmatrix}.$$

As the algorithm is often described, we subtract multiples of the first row from the other rows to eliminate the components of $a_{21}$. In matrix terms, we can describe this process as follows. Let

$$\ell_{21} = \alpha^{-1} a_{21}.$$

Then it is easily verified that

$$\begin{pmatrix} 1 & 0 \\ -\ell_{21} & I \end{pmatrix} \begin{pmatrix} \alpha_{11} & a_{12}^{\mathrm{T}} \\ a_{21} & A_{22} \end{pmatrix} = \begin{pmatrix} \alpha_{11} & 0 \\ a_{21} & A_{22} - \alpha^{-1} a_{21} a_{12}^{\mathrm{T}} \end{pmatrix}. \qquad (24.6)$$

The process is then repeated on the matrix

$$A_{22} - \alpha^{-1} a_{21} a_{12}^{\mathrm{T}},$$

which is called the *Schur complement* of $\alpha_{11}$.

11. Our principal result concerning diagonal dominance and Gaussian elimination is the following.

Let $A$ be a nonsingular diagonally dominant matrix. Then the Schur complement of $\alpha_{11}$ is nonsingular and diagonally dominant.

The first thing we must establish is that the Schur complement exists, i.e., that $\alpha_{11} \neq 0$. But if $\alpha_{11} = 0$, by diagonal dominance the entire first row is zero, and $A$ is singular — a contradiction.

To establish the nonsingularity of the Schur complement, note that the right-hand side of (24.6) is nonsingular because it is the product of nonsingular matrices. Since it is block triangular with the Schur complement as one of its blocks, the Schur complement is also nonsingular.

To establish that the Schur complement is diagonally dominant, let

$$A_{22} = D + B,$$

where $D$ consists of the diagonals of $A_{22}$ and $B$ consists of the off-diagonals. If we let $\mathbf{e}$ be the vector of all ones, then one way of saying that $A_{22}$ is diagonally dominant is to write

$$|D|\mathbf{e} - |B|\mathbf{e} \geq 0.$$

Here the absolute value of a matrix $A$ is defined to be the matrix whose elements are $|\alpha_{ij}|$, and the inequality is taken componentwise.

We are going to show that the Schur complement is diagonally dominant by showing that

$$s \equiv |D|\mathbf{e} - |B| - |\alpha_{11}^{-1} a_{21} a_{12}^{\mathrm{T}}|\mathbf{e} \geq 0.$$

Note that we have lumped the diagonals of $|\alpha_{11}^{-1} a_{21} a_{12}|$ with the off-diagonals because they could potentially decrease the diagonal elements of $D$, and decreasing a diagonal element is morally equivalent to increasing a corresponding off-diagonal element.

First we have

$$s \geq |D|\mathbf{e} - |B|\mathbf{e} - |\alpha_{11}^{-1}||a_{21}||a_{12}|^{\mathrm{T}}\mathbf{e}.$$

Now by diagonal dominance $|a_{12}|^{\mathrm{T}}\mathbf{e} \leq |\alpha_{11}|$, or $|\alpha_{11}^{-1}||a_{12}|^{\mathrm{T}}\mathbf{e} \leq 1$. It follows that

$$s \geq |D|\mathbf{e} - |B|\mathbf{e} - |a_{21}|.$$

But since the last $n - 1$ row rows of $A$ are diagonally dominant, the vector on the right-hand side of this inequality is nonnegative. Thus $s \geq 0$, which establishes the result.

12. This result immediately implies that Gaussian elimination goes to completion when it is applied to a nonsingular diagonally dominant matrix. A similar manipulation gives an inequality that can be used to show that the algorithm is numerically stable. We do not give it here, not because it is difficult, but because to interpret it requires the rounding-error analysis of Gaussian elimination as a background.

13. The course of Gaussian elimination is not affected by scaling of the form

$$A \leftarrow D_1 A D_2,$$

where $D_1$ and $D_2$ are nonsingular diagonal matrices.[30] Thus Gaussian elimination goes to completion for any nonsingular matrix that can be scaled to be diagonally dominant.

An important class of matrices that can be so scaled is the class of M-matrices. These matrices have nonpositive off-diagonal elements and nonnegative inverses; i.e.,

---

[30] Warning: We are talking about the unpivoted algorithm. Pivoted versions which move other elements into the diagonals before performing elimination steps *are* affected by diagonal scaling.

1. $i \neq j \implies \alpha_{ij} \leq 0$,
2. $A^{-1} \geq 0$.

It can be shown that if $A$ is an M-matrix, then there is a diagonal matrix $D$ such that $D^{-1}AD$ is strictly diagonally dominant. Hence the results on Gaussian elimination apply to M-matrices as well as to nonsingular diagonally dominant matrices.

## Return to incomplete factorization

14. We now consider the existence of incomplete LU factorizations.

> If $A$ is a nonsingular diagonally dominant matrix, then $A$ has an incomplete LU decomposition for any sparsity set $\mathcal{Z}$.

To establish the result, we will show that the algorithm in Figure 23.1 amounts to performing Gaussian elimination on a sequence of diagonally dominant matrices. Partition $A$ in the form

$$A = \begin{pmatrix} \alpha_{11} & a_{12}^{\mathrm{T}} \\ a_{21} & A_{22} \end{pmatrix}.$$

Let $a_{21}$ and $a_{12}^{\mathrm{T}}$ be decomposed in the form

$$a_{21} = \hat{a}_{21} + r_{21} \quad \text{and} \quad a_{12}^{\mathrm{T}} = \hat{a}_{12}^{\mathrm{T}} + r_{12}^{\mathrm{T}},$$

where $\hat{a}_{21}$ and $\hat{a}_{12}^{\mathrm{T}}$ are obtained by setting to zero the elements in the sparsity set. Then $r_{21}$ and $r_{12}^{\mathrm{T}}$ are zero off the sparsity set, and by the result of §24.7

$$\hat{A} = \begin{pmatrix} \alpha_{11} & \hat{a}_{12}^{\mathrm{T}} \\ \hat{a}_{21} & A_{22} \end{pmatrix}$$

is a nonsingular diagonally dominant matrix. Let

$$L_1 = \begin{pmatrix} 1 & 0 \\ \alpha_{11}^{-1}\hat{a}_{21} & 0 \end{pmatrix}$$

and

$$U_1 = \begin{pmatrix} \alpha_{11} & a_{12}^{\mathrm{T}} \\ 0 & 0 \end{pmatrix}.$$

Then it is easy to verify that

$$A - L_1 U_1 = \begin{pmatrix} 0 & r_{12}^{\mathrm{T}} \\ r_{21} & \tilde{A}_{22} \end{pmatrix},$$

where

$$\tilde{A}_{22} = A_{22} - \alpha_{11}^{-1}\hat{a}_{21}\hat{a}_{12}^{\mathrm{T}}.$$

Now $\tilde{A}_{22}$ is the Schur complement of $\alpha_{11}$ in $\hat{A}$ and hence is a nonsingular diagonally dominant matrix. Thus, by induction, it has an incomplete factorization

$$\tilde{A}_{22} = L_{22}U_{22} + R_{22}.$$

If we set

$$L = \begin{pmatrix} 1 & 0 \\ \alpha_{11}^{-1}\hat{a}_{21} & L_{22} \end{pmatrix}$$

and

$$U = \begin{pmatrix} \alpha_{11} & \hat{a}_{12}^{\mathrm{T}} \\ 0 & U_{22} \end{pmatrix},$$

then it is easy to verify that

$$A - LU = \begin{pmatrix} 0 & r_{12}^{\mathrm{T}} \\ r_{21} & R_{22} \end{pmatrix} \equiv R.$$

Thus $L$, $U$, and $R$ exhibit the right sparsity patterns, so that $LU$ is an incomplete LU factorization of $A$.

15.    By the comments in §24.13 the incomplete LU factorization also exists for any M-matrix.

- Iterations, Linear and Nonlinear

# Iterations, Linear and Nonlinear

Some Classical Iterations
Splittings and Iterative Methods
Convergence
Irreducibility
Splittings of Irreducibly Diagonally Dominant Matrices
M-Matrices and Positive Definite Matrices

## Some classical iterations

1. In this lecture we will briefly treat some classical iterative methods. Although conjugate gradients, with appropriate preconditioning, has largely replaced the older iterations for solving positive definite systems, the classical methods are worth studying for two reasons. First, they can be applied to nonsymmetric systems. Second, some of the classical methods can be used as preconditioners.

The theory of classical iterative methods is a book in itself and cannot be presented in a single lecture. We will therefore confine ourselves to analyzing methods for diagonally dominant systems. This special case is not only useful but also representative of many techniques used in analyzing classical iterative methods.

2. Consider the linear system

$$Ax = b.$$

If we solve the $i$th equation for the $i$th component of $x$, we get

$$x_i = a_{ii}^{-1}(b_i - \textstyle\sum_{j \neq i} a_{ij} x_j).$$

The form of this expression suggests substituting an approximate solution $x^k$ into it to get an "improved solution"

$$x_i^{k+1} = a_{ii}^{-1}(b_i - \textstyle\sum_{j \neq i} a_{ij} x_j^k). \tag{25.1}$$

If we repeat this procedure for $i = 1, \ldots, n$ we obtain a new vector $x^{k+1}$. We can, of course, repeat the procedure iteratively to generate a sequence of vectors $x^0, x^1, \ldots$.[31] The iteration is called *Jacobi's iteration*.

3. In passing from $x^k$ to $x^{k+1}$ we have to compute the components of $x^{k+1}$ in some order — here assumed to be $1, 2, \ldots, n$. Now, you might ask: If $x_i^{k+1}$ is

---

[31] The use of a superscript — e.g., $x^k$ — to indicate iteration is useful when one must repeatedly refer to the components of a vector — e.g., $x_i^k$. It widely used in the optimization and nonlinear equations literature.

truly better that $x_i^k$, shouldn't we use it in forming the subsequent components of $x^{k+1}$? In other words, why not iterate according to the formulas

$$x_i^{k+1} = a_{ii}^{-1}(b_i - \sum_{j<i} a_{ij} x_j^{k+1} - \sum_{j>i} a_{ij} x_j^k)? \qquad (25.2)$$

It turns out that this method — the *Gauss–Seidel method* — is often faster than Jacobi's method.

4. An important generalization of the Gauss–Seidel method can be obtained by partitioning the equation $Ax = b$ in the form

$$\begin{pmatrix} A_{11} & A_{12} & \cdots & A_{1,k-1} & A_{1k} \\ A_{21} & A_{22} & \cdots & A_{2,k-1} & A_{2k} \\ \vdots & \vdots & & \vdots & \vdots \\ A_{k-1,1} & A_{k-1,2} & \cdots & A_{k-1,k-1} & A_{k-1,k} \\ A_{k1} & A_{k2} & \cdots & A_{k,k-1} & A_{kk} \end{pmatrix} \begin{pmatrix} x_1 \\ x_2 \\ \vdots \\ x_{k-1} \\ x_k \end{pmatrix} = \begin{pmatrix} b_1 \\ b_2 \\ \vdots \\ b_{k-1} \\ b_k \end{pmatrix},$$

where the diagonal blocks $A_{ii}$ are nonsingular. Then in analogy with (25.2), we can write

$$x_i^{k+1} = A_{ii}^{-1}(b_i - \sum_{j<i} A_{ij} x_j^{k+1} - \sum_{j>i} A_{ij} x_j^k).$$

This is the *block Gauss–Seidel method.*

5. The essentially scalar character of the above formulas makes it difficult to analyze the methods. To get a nicer formulation write

$$A = D - L - U,$$

where $D$ consists of the diagonal elements of $A$, and $L$ and $U$ are the subdiagonal and superdiagonal triangles of $A$. Then the sums $\sum_{j\neq i} a_{ij} x_j^k$ in (25.1) are the components of the vector $-(L + U)x^k$. Similarly, multiplication by $a_{ii}^{-1}$ corresponds to multiplication by $D^{-1}$. Thus we can represent the Jacobi iteration in the form

$$x^{k+1} = D^{-1}b + D^{-1}(L + U)x^k.$$

6. For the Gauss–Seidel iteration, the sum $\sum_{j<i} a_{ij} x_j^{k+1}$ written in matrix terms becomes $-Lx^{k+1}$, and the sum $\sum_{j>i} a_{ij} x_j^k$ becomes $-Ux^k$. Thus the Gauss–Seidel iteration assumes the form

$$x^{k+1} = D^{-1}(b + Lx^{k+1} + Ux^k).$$

With a little algebra, we can put this relation in the form

$$x^{k+1} = (D - L)^{-1}b + (D - L)^{-1}Ux^k. \qquad (25.3)$$

7. The block Gauss–Seidel iteration assumes the form (25.3) if we set

$$D = \text{diag}(A_{11}, \ldots, A_{kk}), \tag{25.4}$$

$$L = -\begin{pmatrix} 0 & 0 & \cdots & 0 & 0 \\ A_{21} & 0 & \cdots & 0 & 0 \\ \vdots & \vdots & & \vdots & \vdots \\ A_{k-1,1} & A_{k-1,2} & \cdots & 0 & 0 \\ A_{k1} & A_{k2} & \cdots & A_{k,k-1} & 0 \end{pmatrix}, \tag{25.5}$$

and

$$U = -\begin{pmatrix} 0 & A_{12} & \cdots & A_{1,k-1} & A_{1k} \\ 0 & 0 & \cdots & A_{2,k-1} & A_{2k} \\ \vdots & \vdots & & \vdots & \vdots \\ 0 & 0 & \cdots & 0 & A_{k-1,k} \\ 0 & 0 & \cdots & 0 & 0 \end{pmatrix}. \tag{25.6}$$

## Splittings and iterative methods

8. The three iterations share a common form. To see this, let $A$ be written in the form

$$A = M - N,$$

where $M$ is nonsingular. We call such an expression a *splitting of $A$*.

9. The equation $Ax = b$ can be written in the form $(M - N)x = b$, or equivalently

$$x = M^{-1}b + M^{-1}Nx.$$

This expression suggests an iterative scheme of the form

$$x^{k+1} = M^{-1}b + M^{-1}Nx^k.$$

It is now easy to see that the Jacobi iteration corresponds to a splitting of the form

$$A = D - (L + U).$$

On the other hand, the Gauss–Seidel iteration corresponds to the splitting

$$A = (D - L) - U.$$

The block Gauss–Seidel iteration corresponds to a splitting of the same form but with $D$, $L$, and $U$ defined by (25.4), (25.5), and (25.4).

## Convergence

10.  It is now time to investigate the convergence of the iteration

$$x^{k+1} = M^{-1}b + M^{-1}Nx^k. \qquad (25.7)$$

The first thing to note is that the solution $x$ is a fixed point of the iteration; i.e.,

$$x = M^{-1}b + M^{-1}Nx. \qquad (25.8)$$

If we subtract (25.8) from (25.7), we obtain the following recursion for the error in the iterates:

$$x^{k+1} - x = M^{-1}N(x^k - x).$$

The solution of this recursion is

$$x^k - x = (M^{-1}N)^k(x^0 - x).$$

Consequently, a sufficient condition for the convergence of the method is that

$$\lim_{k \to \infty} (M^{-1}N)^k = 0. \qquad (25.9)$$

By the results of §14.14, a necessary and sufficient for (25.9) is that

$$\rho(M^{-1}N) < 1,$$

where $\rho(M^{-1}N)$ is the spectral radius of $A$.

11.  The condition that $\rho(M^{-1}N) < 1$ is sufficient for the convergence of the iteration (25.7). It is not quite necessary. For example, if $x^0 - x$ is an eigenvector of $M^{-1}N$ corresponding to an eigenvalue whose absolute values is less than one, then the iteration will converge. As a practical matter, however, rounding error will introduce components along eigenvectors corresponding to the eigenvalues greater than or equal to one, and the iteration will fail to converge. For this reason we are not much interested in splittings for which $\rho(M^{-1}N) \geq 1$.

12.  We must now determine conditions under which $\rho(M^{-1}N) < 1$. The solution of this problem depends on the structure of $A$. As we said at the beginning of the lecture, we are going to consider splitting of diagonally dominant matrices. However, it turns out that diagonal dominance alone is not enough to do the job. We also need the concept of irreducibility, to which we now turn.

## Irreducibility

13.  In §24.7 we used row and column interchanges to order the diagonals of a diagonally dominant matrix. More generally, suppose we want the diagonal elements of $A$ to appear in order $k_1, \ldots, k_n$. Let

$$P = (\mathbf{e}_{k_1} \ \cdots \ \mathbf{e}_{k_n}),$$

where $\mathbf{e}_i$ is the vector with $i$th component one and its other components zero. (The matrix $P$ is called a *permutation matrix*.) It is easy to verify that the $(i,j)$-element of $P^{\mathrm{T}}AP$ is $a_{k_i k_j}$. In particular, the $i$th row of $P^{\mathrm{T}}AP$ has the same elements as the $k_i$th row of $A$, so that diagonal dominance is preserved.

14.    We now say that $A$ is *reducible* if there is a permutation matrix $P$ such that

$$P^{\mathrm{T}}AP = \begin{pmatrix} A_{11} & 0 \\ A_{21} & A_{22} \end{pmatrix}, \tag{25.10}$$

where $A_{11}$ is square. Otherwise $A$ is *irreducible*. We have already seen a reducible matrix in (24.5).

15.    Although reducibility or irreducibility is not easily determined from the definition, there is an alternative characterization that is easy to apply. Let us say that you can "step" from the diagonal element $a_{ii}$ to $a_{jj}$ if $a_{ij} \neq 0$. Then $A$ is irreducible if and only if you can travel from any diagonal element to any other by a sequence of steps. In using this result, it is convenient to put down $n$ dots representing the diagonals and connect them with arrows representing permissible steps.[32]

This result is a little tricky to prove, but you can easily convince yourself that in the reducible matrix (25.10) there is no way to travel from a diagonal element of $A_{11}$ to a diagonal element of $A_{22}$.

16. We will now couple irreducibility with diagonal dominance in the following definition. A matrix $A$ is said to be *irreducibly diagonally dominant* if

1.  $A$ is irreducible,
2.  $A$ is diagonally dominant with strict inequality in at least one row.

Note that the strict-inequality requirement is not reflected in the term "irreducibly diagonally dominant" — sometimes a source of confusion.

## Splittings of irreducibly diagonally dominant matrices

17.    The principal result of this division is the following.

> Let $A$ be irreducibly diagonally dominant, and let $A = M - N$ be a splitting such that
>
> 1.  The diagonal elements of $N$ are zero,
> 2.  $|A| = |M| + |N|$.
>
> Then $M$ is nonsingular and $\rho(M^{-1}N) < 1$.

$\qquad\qquad\qquad\qquad\qquad\qquad\qquad\qquad\qquad\qquad\qquad\qquad\qquad\qquad$ (25.11)

---

[32]This structure is called the *graph of A*, and it is useful in many applications.

Here $|A|$ is the matrix obtained from $A$ by taking the absolute value of its elements.

18. Of the two conditions, the first simply says that the diagonals of $A$ and $M$ are the same. The second condition, written elementwise in the form $|a_{ij}| = |m_{ij}| + |n_{ij}|$, says that the elements of $M$ are obtained by reducing the elements of $A$ in magnitude but without changing their signs.

19. To prove this result, first note that by the result of §24.7 $M$ is nonsingular. The proof that $\rho(M^{-1}N) > 1$ is by contradiction. Assume that

$$M^{-1}Nx = \lambda x, \qquad x \neq 0, \quad |\lambda| \geq 1.$$

Let $x_i$ be a maximal component of $x$, i.e., a component whose absolute value is largest. Then from the equation $Mx = \lambda^{-1}Nx$, we get

$$a_{ii}x^i + \sum_{j \neq i}(m_{ij} - \lambda^{-1}n_{ij})x^j = 0.$$

(Here we use the fact that $m_{ii} = a_{ii}$.) From this it follows that

$$\begin{aligned}
|a_{ii}| &\leq \sum_{j \neq i}(|m_{ij}| - |\lambda^{-1}||n_{ij}|)\frac{|x_j|}{|x_i|} \\
&\leq \sum_{j \neq i}(|m_{ij}| - |n_{ij}|)\frac{|x_j|}{|x_i|} \\
&\leq \sum_{j \neq i}|a_{ij}|\frac{|x_j|}{|x_i|} \\
&\leq \sum_{j \neq i}|a_{ij}|.
\end{aligned} \qquad (25.12)$$

Now if the inequality is strict, we contradict the diagonal dominance of $A$. If equality holds, then $a_{ij} = 0$ whenever $x_j$ is not maximal [see (24.3)].

Suppose equality holds for all maximal components (otherwise we have our contradiction). Now there must be at least one component that is not maximal; for otherwise $A$ would be diagonally dominant with equality in all row sums. Let $P$ be a permutation that moves the maximal components to the beginning. Partition

$$P^{\mathrm{T}}AP = \begin{pmatrix} A_{11} & A_{12} \\ A_{21} & A_{22} \end{pmatrix},$$

where $A_{11}$ is square and associated with the maximal components. Since the elements of $A_{12}$ are associated with nonmaximal components in (25.12), they are zero. It follows that $A$ is reducible — a contradiction.

20. This is a fairly powerful result, in that it says that any reasonable splitting of an irreducibly diagonally dominant matrix defines a convergent iteration. In particular, the Jacobi, Gauss–Seidel, and block Gauss–Seidel iterations all

converge. Unfortunately, the result does not say which is the best. But if I were forced to bet, I would choose block Gauss–Seidel and would probably win.

21.   The irreducibility hypothesis was used to argue to a contradiction when equality held in all the sums (25.12). But if we assume that $A$ is strictly diagonally dominant, even equality is a contradiction. Hence we have the following result.

> The result of §25.17 remains true if irreducible diagonal dominance is replaced by strict diagonal dominance.

## M-matrices and positive definite matrices

22.   Recall that an M-matrix is a nonsingular matrix with nonpositive off diagonals and nonnegative inverse (see §24.13). If $A$ is an M-matrix, then there is a diagonal matrix $D$ such that $D^{-1}AD$ is strictly diagonally dominant. Now if $A = M - N$ is a splitting of $A$ that satisfies the conditions (25.11), then the splitting $D^{-1}AD = D^{-1}MD - D^{-1}ND$ satisfies the same conditions. Moreover, the iteration matrix $(D^{-1}MD)^{-1}(D^{-1}ND) = D^{-1}(M^{-1}N)D$ is similar to $M^{-1}N$ and therefore has the same spectral radius. Thus our result applies to $M$ matrices, reducible or irreducible.

However, we can prove far more for M-matrices. For example, we can compare rates of convergence. In fact, we can derive our results for diagonally dominant matrices from the theory of M-matrices. But that is another story.

23.   As we have mentioned, positive definite systems are usually treated by conjugate gradients. However, their splittings have nice properties. We state the following result without proof.

> Let $A$ be positive definite and let
>
> $$A = P - L - L^{\mathrm{T}},$$
>
> where $P$ is positive definite and $P - L$ is nonsingular. Then
>
> $$\rho[(P - L)^{-1}L^{\mathrm{T}}] < 1.$$

This result implies that block Gauss–Seidel for positive definite matrices always converges.

# Iterations, Linear and Nonlinear

Linear and Nonlinear
Continuity and Derivatives
The Fixed-Point Iteration
Linear and Nonlinear Iterations Compared
Rates of Convergence
Newton's Method
The Contractive Mapping Theorem

## Linear and nonlinear

1.  In the last lecture we considered the linear iteration

$$x^{k+1} = (M^{-1}N)x^k + M^{-1}b$$

for solving the system $Ax = b$. If we set

$$g(x) = (M^{-1}N)x + M^{-1}b,$$

we can write this iteration in the form

$$x^{k+1} = g(x^k). \tag{26.1}$$

The function $g$ is called an *iteration function.*
   The solution $x_*$ of the system satisfies

$$x_* = g(x_*);$$

i.e., it remains invariant under the action of $g$. For this reason it is called a *fixed point* of $g$, and the iteration (26.1) is called a *fixed-point iteration.* We know that the iteration converges from any starting point $x^0$, provided

$$\rho(M^{-1}N) < 1.$$

2.  In this lecture, we will investigate what happens when $g$ is a nonlinear function mapping some subset of $\mathbf{R}^n$ into $\mathbf{R}^n$. Specifically, we will assume that $g$ has a fixed point $x_*$ and ask under what conditions the iteration $x^{k+1} = g(x^k)$ converges to $x_*$ from a starting point $x^0$.

## Continuity and derivatives

3.  Our analysis will require that we extend the notion of derivative to $R^n$. Fortunately, if we are careful about the way we define things, we can get away with a minimal amount of analytic clutter. We will begin with a useful definition of continuity. For the rest of this lecture $\|\cdot\|$ will denote a vector norm and a consistent matrix norm.

4.  We will say that a function $f: \mathbf{R}^n \to \mathbf{R}^n$ is continuous at a point $x_*$ if

$$f(x) = f(x_*) + q(x), \tag{26.2}$$

where

$$\lim_{x \to x_*} \|q(x)\| = 0. \tag{26.3}$$

This definition immediately implies that $\lim_{x \to x_*} \|f(x) - f(x_*)\| = 0$, or

$$\lim_{x \to x_*} f(x) = f(x_*),$$

which is one of the usual definitions of continuity.

A convenient abbreviation for (26.2) and (26.3) is

$$f(x) = f(x_*) + o(1),$$

where $o(1)$ is a generic function that approaches zero as $x \to x_*$.

5.  In defining the derivative of a multivariate function, the last thing we want is to get involved with partial derivatives, chain rules, and the like. Instead we will characterize the derivative as a matrix with certain properties. The necessary properties can be seen by considering the case of one variable. If we expand $f(x)$ in a Taylor series, we get

$$f(x) = f(x_*) + f'(x_*)(x - x_*) + r(x), \tag{26.4}$$

where[33]

$$\|r(x)\| \le \epsilon(\|x - x_*\|)\|x - x_*\|, \tag{26.5}$$

for some function $\epsilon(t)$ satisfying

$$\lim_{t \to 0} \epsilon(t) = 0. \tag{26.6}$$

Note that these equations imply that

$$f'(x_*) = \lim_{x \to x_*} \frac{f(x) - f(x_*)}{x - x_*},$$

which is the usual definition of the derivative.

---

[33] We use norms rather than absolute values in anticipation of the multivariate case.

If $f$ is a mapping from $\mathbf{R}^n$ to $\mathbf{R}^n$, the expression (26.4) makes sense only if $f'(x)$ is a matrix of order $n$. We will therefore say that $f$ is differentiable at $x_*$ if there is a matrix $f'(x_*)$ such that (26.4), (26.5), and (26.6) are satisfied. It is easy to verify that $f'$ is the matrix whose $(i,j)$-element is the partial derivative of the $i$th component of $f$ with respect to the $j$th component of $x$. It is also called the *Jacobian* of $f$.

We will abbreviate (26.5) and (26.6) by writing

$$r(x) = o(x - x_*),$$

where $o(x - x_*)$ is a generic function that goes to zero faster than $x - x_*$.

## The fixed-point iteration

6. We are now in a position to analyze the iteration $x^{k+1} = g(x^k)$. We assume that $g$ is differentiable at the fixed point $x_*$, so that

$$g(x) = g(x_*) + g'(x_*)(x - x_*) + r(x),$$

where $r(x) = o(x - x_*)$. Since $g(x_*) = x_*$, we have

$$g(x) = x_* + g'(x_*)(x - x_*) + r(x - x_*).$$

In particular, $x^{k+1} = g(x^k) = x_* + g'(x_*)(x^k - x_*) + r(x^k - x_*)$, or

$$x^{k+1} - x_* = g'(x_*)(x^k - x_*) + r(x^k - x_*).$$

Let us now take norms to get

$$\|x^{k+1} - x_*\| \le \|g'(x_*)\|\|(x^k - x_*)\| + \|r(x^k - x_*)\|.$$

By (26.5),

$$\|x^{k+1} - x_*\| \le (\|g'(x_*)\| + \epsilon(\|x^k - x_*\|)\|(x^k - x_*)\|, \qquad (26.7)$$

where $\epsilon = o(1)$.

7. The inequality (26.7) suggests that we will not have much success in proving convergence if $\|g'(x_*)\| > 1$. For in that case, as $x^k$ approaches $x_*$, the term $\epsilon(\|x^k - x_*\|)$ becomes negligible, and the bound represents a potential increase in the error in $x^{k+1}$ over that of $x^k$. Hence we will assume that

$$\|g'(x_*)\| < 1.$$

8. The inequality (26.7) also suggests that we must restrict the starting point $x_0$ to be in some sense near $x_*$. For if $x^0$ is not near $x_*$, we cannot say anything

about the size of $\epsilon(\|x^k - x_*\|)$. We will make the following restriction. Because $\epsilon(\|x^k - x_*\|) = o(1)$, there is a $\delta > 0$ such that

$$\|x - x_*\| \leq \delta \implies \|g'(x_*)\| + \epsilon(\|x^k - x_*\|) \leq \alpha < 1.$$

We will require that

$$\|x^0 - x_*\| \leq \delta.$$

9. With these assumptions, it follows from (26.7) that

$$\|x^1 - x_*\| \leq \alpha\|x^0 - x_*\|.$$

This implies that $\|x^1 - x_*\| \leq \delta$. Hence

$$\|x^2 - x_*\| \leq \alpha\|x^1 - x_*\| \leq \alpha^2\|x^0 - x_*\|.$$

In general

$$\|x^k - x_*\| \leq \alpha^k\|x^0 - x_*\|.$$

and since $\alpha < 1$, the $x^k$ converge to $x_*$.

10. The analysis suggests that the iteration converges at least as fast as $\alpha^k$ approaches zero. However, $\alpha$ has been inflated to account for the behavior of $g(x)$ when $x$ is away from $x_*$. A more realistic assessment of the rate of convergence may be obtained by writing (26.7) in the form

$$\frac{\|x^{k+1} - x_*\|}{\|x^k - x_*\|} \leq \|g'(x_*)\| + \epsilon(\|x^k - x_*\|).$$

Since $\epsilon = o(1)$,

$$\lim_{k \to \infty} \frac{\|x^{k+1} - x_*\|}{\|x^k - x_*\|} \leq \|g'(x_*)\|.$$

Hence

$$\|x^{k+1} - x_*\| \lesssim \|g'(x_*)\|\|x^k - x_*\|.$$

In other words, the error in $x^{k+1}$ is asymptotically less than the error in $x^k$ by a factor of at least $\|g'(x_*)\|$.

11. We may summarize these results as follows.

> Let $g$ be differentiable at a fixed point $x_*$. If
>
> $$\|g'(x_*)\| < 1, \tag{26.8}$$
>
> then there is a $\delta > 0$ such that for any starting point $x^0$ satisfying $\|x^0 - x_*\| \le \delta$ the iteration
>
> $$x^{k+1} = g(x^k), \qquad k = 0, 1, \ldots$$
>
> converges to $x_*$. Moreover,
>
> $$\lim_{k\to\infty} \frac{\|x^{k+1} - x_*\|}{\|x^k - x_*\|} \le \|g'(x_*)\|.$$

## Linear and nonlinear iterations compared

12. It is instructive to compare these results with those for the linear iteration

$$x^{k+1} = (M^{-1}N)x^k + M^{-1}b \equiv g(x^k).$$

The condition for the convergence of this iteration is $\rho(M^{-1}N) < 1$. But in this case the derivative of $g$ is identically $M^{-1}N$. Since the norm of a matrix bounds its spectral radius, the condition (26.8) insures that the iteration converges.[34]

13. The nonlinear iteration differs from the linear iteration in an important aspect. The linear iteration is *globally convergent* — it converges from any starting point $x^0$. The nonlinear iteration is, in general, only *locally convergent* — the starting point must be sufficiently near the fixed point to insure convergence.

## Rates of convergence

14. A converging sequence for which

$$\frac{\|x^{k+1} - x_*\|}{\|x^k - x_*\|} \to \alpha < 1$$

is said to *converge linearly*. A fixed-point iteration usually converges linearly, and $\alpha$ will be the spectral radius of $g'(x_*)$. An important exception is when $g'(x_*) = 0$, in which case the convergence is said to be *superlinear*. In this case, the convergence accelerates as the iteration progresses.

15. A particularly important case of superlinear convergence occurs when $g'(x_*) = 0$ and $\epsilon(\|x^k - x_*\|) \le K\|x^k - x_*\|$ for some constant $K$. In this case

$$\|x^{k+1} - x_*\| \le K\|x^k - x_*\|^2.$$

---

[34] In fact, we could replace (26.8) by the condition $\rho[g'(x_*)] < 1$. For by the result of §14.12, if $\rho[g'(x_*)] < 1$, there is a norm such that $\|g'(x_*)\| < 1$.

In particular, if

$$\|x^{k+1} - x_*\| \cong K\|x^k - x_*\|^2,$$

then we say the sequence *converges quadratically*. We have encountered quadratic convergence in the nonsymmetric QR algorithm (see §15.20).

## Newton's method

16. We are now going to apply our analysis of the fixed-point iteration to analyze the multivariate version of Newton's method. To derive the algorithm, suppose $f: \mathbf{R}^n \to \mathbf{R}^n$, and let $f(x_*) = 0$. Let us assume that $f'(x)$ is continuous and nonsingular about $x_*$. Then

$$0 = f(x_*) = f(x) + f'(x)(x_* - x) + o(x_* - x) = f'(x_*)(x_* - x) + o(x_* - x).$$

If we drop the $o(x_* - x)$ term and solve for $x_*$, we get

$$x_* \cong x - f'(x)^{-1}f(x).$$

The multivariate form of Newton's method is obtained by iterating this formula starting with some initial approximation $x^0$ to $x_*$:

$$x^{k+1} = x^k - f'(x^k)^{-1}f(x^k), \qquad k = 0, 1, \dots.$$

17. The iteration function for Newton's method is

$$g(x) = x - f'(x)^{-1}f(x). \tag{26.9}$$

Because $f(x_*) = 0$, we have $g(x_*) = x_*$; i.e., $x_*$ is a fixed point of $g$. According to our local convergence theory, the behavior of Newton's method depends on the size of $g'(x_*)$.

In the univariate case, we can evaluate $g'$ using elementary differentiation formulas. We can do something like that in the multivariate case; however, the formulas are not simple. Instead we will compute the first two terms of the Taylor series for $g(x)$ about $x_*$. Whatever multiplies $x - x_*$ must be the required derivative.

Since $f$ is differentiable at $x_*$ and $f(x^*) = 0$,

$$f(x) = f(x_*) + f'(x_*)(x - x_*) + o(x - x_*) = f'(x_*)(x - x_*) + o(x - x_*). \tag{26.10}$$

Since $f'(x)$ is continuous,

$$f'(x) = f'(x_*) + o(1).$$

Since the inverse of a matrix is continuous,

$$f'(x)^{-1} = [f'(x_*) + o(1)]^{-1} = f'(x_*)^{-1} + o(1). \tag{26.11}$$

If we substitute (26.10) and (26.11) into (26.9), we get

$$\begin{aligned}
g(x) &= x - [f'(x_*)^{-1} + o(1)][f'(x_*)(x - x_*) + o(x - x_*)] \\
&= x - (x - x_*) - f'(x_*)o(x - x_*) - o(1)f'(x_*)(x - x_*) - o(1)o(x - x_*) \\
&= x_* + o(x - x_*) \\
&= g(x_*) + 0 \cdot (x - x_*) + o(x - x_*).
\end{aligned}$$
$$(26.12)$$

It follows that $g'(x_*) = 0$.

18.  Applying our fixed-point theory and our observations on superlinear convergence, we get the following result.

> Let $f(x_*) = 0$ and suppose that $f'(x)$ is continuous at $x_*$. Then for all $x^0$ sufficiently near $x_*$, the Newton iteration
>
> $$x^{k+1} = x^k - f'(x^i)^{-1} f(x^k), \qquad k = 0, 1, \ldots$$
>
> converges superlinearly to $x_*$. If $f''(x_*)$ exists, the convergence is at least quadratic.

The result on quadratic convergence follows from the fact that if $f$ is twice differentiable at $x_*$ then the norm of $o(1)$ in (26.12) can be bounded by a multiple of $\|x - x_*\|$ and the norm of $o(x - x_*)$ can be bounded by a multiple of $\|x - x_*\|^2$.

## The contractive mapping theorem

19.  One of the difficulties with the fixed-point iteration is that we have to know of the existence of a fixed point and we have to position $x^0$ sufficiently near it. In some instances, however, we can guarantee the existence of a fixed point. Here we will turn to one example.

20.  A function $g$ is a *contraction* if there is a constant $\alpha < 1$ such that

$$\|g(x) - g(y)\| \leq \alpha \|x - y\|. \qquad (26.13)$$

In other words the images of two points under a contraction are nearer than the original points.

21.  A contraction is uniformly continuous. For by (26.13), if $\epsilon > 0$, then $\|g(x) - g(y)\| \leq \epsilon$ whenever $\|x - y\| \leq \epsilon$.

22.  We are going to show that under certain circumstances a contractive mapping has a unique fixed point and that fixed-point iteration converges to it from any starting point. To establish these facts, we will use a result on the convergence of sequences in $\mathbf{R}^n$ that is of independent interest. We state it without proof.

> Let $x^0, x^1, \ldots$ be a sequence of vectors in $\mathbf{R}^n$. If there are constants $\beta_0, \beta_1, \ldots$ satisfying
>
> $$\begin{aligned} 1. \quad & \|x^{k+1} - x^k\| \leq \beta_k, \\ 2. \quad & \textstyle\sum_{k=0}^{\infty} \beta_k < \infty, \end{aligned}$$
>
> then there is an $x_* \in \mathbf{R}^n$ such that $\lim_{k\to\infty} x^k = x_*$. Moreover,
>
> $$\|x_0 - x_*\| \leq \sum_{k=0}^{\infty} \beta_k.$$

23.   We are now in a position to state and prove the contractive mapping theorem.

> Let $\mathcal{C}$ be a closed subset of $\mathbf{R}^n$ and let $g \colon \mathcal{C} \to \mathbf{R}^n$ be a contraction with constant $\alpha$. If
>
> $$x \in \mathcal{C} \implies g(x) \in \mathcal{C},$$
>
> then $g$ has a unique fixed point $x_*$ in $\mathcal{C}$. Moreover, for any $x^0 \in \mathcal{C}$ the sequence defined by
>
> $$x^{k+1} = g(x^k), \qquad k = 0, 1, \ldots,$$
>
> converges to $x_*$ and
>
> $$\|x^0 - x_*\| \leq \frac{\|x^1 - x^0\|}{1 - \alpha}. \tag{26.14}$$

To establish this result, write

$$\|x^{k+1} - x^k\| = \|g(x^k) - g(x^{k-1})\| \leq \alpha\|x^k - x^{k-1}\|.$$

Hence by an obvious induction

$$\|x^{k+1} - x^k\| \leq \alpha^k\|x^1 - x^0\| \equiv \beta_k.$$

Now

$$\sum_{k=0}^{\infty} \beta_k = \|x^1 - x^0\| \sum_{k=0}^{\infty} \alpha^k = \frac{\|x^1 - x^0\|}{1 - \alpha}.$$

Hence by the result of §26.22, the $x^k$ have a limit $x_*$ that satisfies (26.14).

To show that $x_*$ is a fixed point of $g$, note that because $\mathcal{C}$ is closed $x_* \in \mathcal{C}$. Hence by the continuity of $g$,

$$x_* = \lim_{k\to\infty} x^k = \lim_{k\to\infty} g(x^{k-1}) = g(x_*).$$

Finally $x_*$ is unique. For if $x_+$ is any other fixed point of $g$,

$$\|x_* - x_+\| = \|g(x_*) - g(x_+)\| \le \alpha\|x_* - x_+\|,$$

which is impossible unless $x_* = x_+$.

24. Although the contractive mapping theorem has its uses, it is often difficult to apply in practice. The usual problem is to find the set $\mathcal{C}$ which is mapped into itself. A trick that often works is to mimic the proof of the theorem by starting with a particular point and bounding the norms of the differences in the iterates. If sum of the bounds converges, you have your fixed point. Here, as with so many constructive results in numerical analysis, the proof is more useful than the result itself.

# Bibliography

## Introduction

1. This bibliography consists of texts and monographs that treat and expand the material in these Afternotes. Although most of the topics have settled down, the subject of Krylov methods for nonsymmetric eigenvalue problems is currently in a state of flux, and no definitive treatment exists. I have instead referenced a report, available over the web, for a package implementing the iteratively restarted Arnoldi method.

## Bibliography

E. Anderson, Z. Bai, C. Bischof, J. Demmel, J. Dongarra, J. Du Croz, A. Greenbaum, S. Hammarling, A. McKenney, S. Ostrouchov, and D. Sorensen. *LAPACK Users' Guide.* SIAM, Philadelphia, second edition, 1995.

K. E. Atkinson. *An Introduction to Numerical Analysis.* John Wiley, New York, 1978.

O. Axelsson. *Iterative Solution Methods.* Cambridge University Press, Cambridge, 1994.

Å. Björck. *Numerical Methods for Least Squares Problems.* SIAM, Philadelphia, 1996.

G. Dahlquist and Å. Björck. *Numerical Methods.* Prentice–Hall, Englewood Cliffs, NJ, 1974.

B. N. Datta. *Numerical Linear Algebra and Applications.* Brooks/Cole, Pacific Grove, CA, 1995.

P. J. Davis. *Interpolation and Approximation.* Blaisdell, New York, 1961. Reprinted by Dover, New York, 1975.

C. de Boor. *A Practical Guide to Splines.* Springer-Verlag, Berlin, 1978.

J. E. Dennis and R. B. Schnabel. *Numerical Methods for Unconstrained Optimization and Nonlinear Equations.* Prentice–Hall, Englewood Cliffs, NJ, 1983.

J. J. Dongarra, J. R. Bunch, C. B. Moler, and G. W. Stewart. *LINPACK User's Guide.* SIAM, Philadelphia, 1979.

G. Farin. *Curves and Surfaces for Computer Aided Geometric Design: A Practical Guide.* Academic Press, New York, 1988.

G. H. Golub and C. F. Van Loan. *Matrix Computations.* Johns Hopkins University Press, Baltimore, MD, third edition, 1996.

N. J. Higham. *Accuracy and Stability of Numerical Algorithms.* SIAM, Philadelphia, 1996.

A. S. Householder. *The Theory of Matrices in Numerical Analysis.* Dover Publishing, New York. Originally published by Blaisdell, New York, 1964.

E. Issacson and H. B. Keller. *Analysis of Numerical Methods.* John Wiley & Sons, New York, 1966.

D. Kincaid and W. Cheney. *Numerical Analysis: Mathematics of Scientific Computing.* Brooks/Cole, Pacific Grove, CA, 1991.

C. L. Lawson and R. J. Hanson. *Solving Least Squares Problems.* Prentice–Hall, Englewood Cliffs, NJ, 1974. Reissued with a survey on recent developments by SIAM, Philadelphia, 1995.

R. B. Lehoucq, D. C. Sorensen, and C Yang. *ARPACK Users' Guide: Solution of Large Scale Eigenvalue Problems by Implicitly Restarted Arnoldi Methods,* 1997. Available at
http//www.caam.rice.edu/software/ARPACK/index.html

J. M. Ortega. *Numerical Analysis: A Second Course.* Academic Press, New York, 1972.

J. M. Ortega and W. C. Rheinboldt. *Iterative Solution of Nonlinear Equations in Several Variables.* Academic Press, New York, 1970.

B. N. Parlett. *The Symmetric Eigenvalue Problem.* Prentice–Hall, Englewood Cliffs, NJ, 1980.

G. W. Stewart. *Introduction to Matrix Computations.* Academic Press, New York, 1973.

G. W. Stewart. *Afternotes on Numerical Analysis.* SIAM, Philadelphia, 1996.

G. W. Stewart and J.-G. Sun. *Matrix Perturbation Theory.* Academic Press, Boston, 1990.

J. Stoer and R. Bulirsch. *Introduction to Numerical Analysis.* Springer-Verlag, New York, second edition, 1993.

R. S. Varga. *Matrix Iterative Analysis.* Prentice–Hall, Englewood Cliffs, NJ, 1962.

D. S. Watkins. *Fundamentals of Matrix Computations.* John Wiley & Sons, New York, 1991.

J. H. Wilkinson. *The Algebraic Eigenvalue Problem.* Clarendon Press, Oxford, England, 1965.

J. H. Wilkinson and C. Reinsch. *Handbook for Automatic Computation. Vol.* II

*Linear Algebra.* Springer, New York, 1971.

D. M. Young. *Iterative Solution of Large Linear Systems.* Academic Press, New York, 1971.

# Index

Italic page numbers signify a defining entry. The abbreviation *me* indicates there is more information at the main entry for this item. The letter "n" indicates a footnote.